TALKING AND LISTENING IN THE AGE OF MODERNITY

Essays on the history of sound

TALKING AND LISTENING IN THE AGE OF MODERNITY

Essays on the history of sound

Edited by Joy Damousi and Desley Deacon

ANU
THE AUSTRALIAN NATIONAL UNIVERSITY

E PRESS

ANU

E PRESS

Published by ANU E Press
The Australian National University
Canberra ACT 0200, Australia
Email: anuepress@anu.edu.au
This title is also available online at: http://epress.anu.edu.au/tal_citation.html

National Library of Australia
Cataloguing-in-Publication entry

Talking and listening in the age of modernity : essays on
the history of sound.

Includes index.
ISBN 9781921313479 (pbk.)
ISBN 9781921313486 (online)

1. Sounds. 2. Auditory perception - Social aspects -
Australia. 3. Oral communication - Social aspects -
Australia. I. Deacon, Desley. II. Damousi, Joy, 1961- .

302.2

Cover design by ANU E Press
Cover image: Photograph of Doris Auguste Heindorff listening to a gramophone, New
Farm, Brisbane, 1903-1913. John Oxley Library, State Library of Queensland.

Table of Contents

Introduction

Listening to the past

Joy Damousi and Desley Deacon

The ubiquitous sound of gun-fire on the Australian gold-fields; the silence of men in the Citroën factory outside Paris as the giant drop-hammers boomed, the milling machines screamed, the boring machines squealed and the pulleys sighed; Australian schoolchildren reciting in one nation-wide accent; Alfred Deakin's virile oratory; William Tilly's dream of a 'World English'; ideas of what was a 'legitimate' or 'pure' Australian accent; powerful speech, intimate speech and 'bad English' in American talkies—the experience and meaning of these sounds and silences have seemed, until recently, too ineffable to be captured by historians.

Historians of modernity have, until recently, focused almost exclusively on sight—when they have considered the senses at all. A major new interest, however, in the history and anthropology of the senses has begun to pay attention to the ways in which modern life has been shaped by the auditory as much as by the visual. Following the example of Alain Corbin—whose pioneering *Village Bells: Sounds and Meaning in the 19ᵗʰ-Century French Countryside* (1994) argued that historians 'can no longer afford to neglect materials pertaining to auditory perception'—this new scholarship aims to capture the auditory environment of the past in order to provide new insights into historical places, events and processes. One of the most suggestive of these new scholars, Steven Connor (1997), argues, in particular, for 'the compelling importance of the auditory in the cultural, clinical and technological constitution of the modern self'.[1]

This collection of essays on talking and listening in the age of modernity takes up the challenge thrown out by this new scholarship by bringing together some of the major scholars of the history of the auditory in Australia. In taking as its focus talking and listening, it follows the dicta of two of Australia's pioneering historians in this new field. Alan Atkinson, in his 2002 book *The Commonwealth of Speech*, argues that talk needs to be understood better as a hidden foundation of human history. 'Speech,' he points out, 'continues, resilient, unpredictable, always strangely powerful, the medium at the heart of human affairs.' Historians, therefore, 'need to make an effort to think of the past as abuzz with voices'—like those on board the *Bounty*, evoked by Greg Dening in *Mr Bligh's Bad Language*—'fill[ing] their territory with chatter.' Joy Damousi, in her 2005 book, *Freud in the Antipodes: A Cultural History of Psychoanalysis in Australia*, makes a similar argument for listening. 'There was a move,' she observes, 'from learning about the world by looking…to interpreting one's surroundings by listening to the radio, conversing on the telephone and going to the cinema, from the 1920s onwards.' She notes, 'While historians and other social theorists have focused primarily on the visual and its influence in the construction of identity, little attention has been given to the auditory in understanding the "modern" notion of the self.'[2]

The essays gathered together in this volume consider the influence of the auditory in the formation of individual and collective subjectivities; the importance of speech to understandings of individual and collective endeavours; the centrality of voice in marking and eliding difference and in struggles for power; and the significance of radio and film in the formation of modern cultural identities.

In Part I of this collection, Diane Collins and James Donald examine two periods whose intensity of change is linked to aural experience as much as to visual: the 'Roaring Twenties' and what Collins calls 'the roaring decade' of the 1850s Australian gold-fields. Both explore 'aural time': the conceptualisation and periodising of history according, not merely to visual and material changes reflected in documents and artefacts, but to the experience of sound. Collins' chapter argues that there are multiple modernities and that the roar of the gold-fields in 1850s Australia was linked to new material environments and new behaviours in a rupture as significant as that of 1920s Paris and London. Donald examines those 1920s sound-scapes—along with those of Berlin and Vienna—identifying the new sounds associated with electricity, the combustion engine and media and communication technologies such as the telephone, the gramophone, the radio and the cinema. He shows how sound began to be thought about in new ways—as something to be measured, managed and abated through new methods of building, new regulatory measures and new techniques of mechanical reproduction—and how a new aesthetic of sound emerged in the novels, poetry and operas of the period.

In Part II, Alan Atkinson demonstrates the importance of uniform speech training in schools in the Australian colonies in the 1870s and 1880s in moulding 'a nation of ready public speakers' who could form a community of truly democratic people. The continental nationalism that resulted in federation of the colonies in 1901 was, he argued, considered to be manifest in the speech of the people. Marilyn Lake, in turn, examines the brilliant oratory of one of those 'ready public speakers', Alfred Deakin—arguing that, in the world of public men, oratory was the measure of manhood, it was a key weapon in Deakin's political battle for the new nation's independence and central to his demand for British recognition of the rights and equality of Australian manhood.

Peter Kirkpatrick examines the rise and fall of recitation in Australia, not as a folkloric remnant, but as the product of modern technologies of speech and performance. He revisits the broader cultural context in which recitation flourished, outlining three factors which led to its popularity: the growth and spread of elocution, the institutionalisation of recitation in schools, and its professionalisation as a performance art. The chapter offers a glimpse into one of the lost arts of everyday life, but also sheds new light on some of the ways in which poetry was mobilised by modern popular culture.

The essays in Part III by Desley Deacon and Bruce Moore examine new chapters in the history of the Australian accent. Bruce Moore argues that the nineteenth-century Australian accent developed from a process of the levelling and excision of obvious dialectal features, resulting in what commentators called a 'pure' accent—that is, one not marked by provincialisms. Yet, he notes, from the late 1880s, a prescriptive attitude towards Australian vowels and diphthongs developed in which they were being judged against an 'ideal' or 'standard' pronunciation. By the first half of the twentieth century, the Australian accent was regarded as impure, ugly and substandard. Moore attributes this change in attitude to the increasing popularity of elocution and phonics, the exponents of which saw 'purity' as an ideal pronunciation rather than a non-regional one. He argues that 'cultivated' Australian and 'broad' Australian developed in the first half of the twentieth century as extremes on either side of the general Australian of the nineteenth century. Since the 1970s, he points out, the former has lost its power, and broad Australian is rarely heard. Most Australians now speak general Australian—something very similar, Moore argues, to the foundation accent of the 1820s.

Desley Deacon places the search for an ideal English pronunciation in a transnational context when she investigates the strange history of 'World English' or 'Good American Speech' through the story of William Tilly, the Australian schoolteacher who taught speech and phonetics to a generation of New York actors and teachers from his base at Columbia University, from 1918 to 1935. Drawing on the weekly column of Tilly's student Windsor Daggett in *Billboard* from 1922 to 1925, Daggett's articles in the influential *Theatre Arts Monthly* during the 1920s, and the activities of other evangelising students throughout the United States, Deacon notes the similarity of 'World English' to what Bruce Moore calls 'Cultivated Australian' and places the success of Tilly's work in the context of ideals of internationalism of the period.

Examining discussions about Australian speech from the 1920s to the early 1940s, Joy Damousi notes the perceived relationship between speech and the formation of the nation. She demonstrates the ways in which a preoccupation with the Australian accent became a means of exploring Australianness, how popular discussion was at first embedded firmly in the notion of Australia as part of the British Empire, which was replaced gradually with pride in the development of a distinctive Australian sound. Speech was also perceived as an indicator of moral worth at the collective and the individual level, and discussions about accents led invariably to considerations of the Australian national character.

Bruce Johnson reminds us of the role of the voice in defining territory and identity—of the connection between voice and power. He offers a deep historical perspective on the way that the relationship between voice and power has changed, arguing that the voice's challenge to print as a major site of power is one of the keys to understanding modernity. Until the late nineteenth century, he argues, the radius of vocalised space was limited by the presence of the human body. From the 1870s, however, this limitation was transcended by the invention of sound recording and related technologies. These permitted a spatial expansion that ultimately became global and, through sound-storage systems, also produced a temporal enlargement. Johnson points out that the small voice—the local and the domestic voice—now clamours for transnational public space; yet it can be turned against its source through various sound-mixing technologies. Illustrating this modern ambiguity with the local example of the politician Pauline Hanson, he shows how sound technology enabled her rise to power, but also helped undermine that power through the spoof *Pauline Pantsdown*, which remixed her own voice to subvert everything she stood for.

The final part of the collection examines the technologies of radio and film and the history of sound and technology. Bridget Griffen-Foley explores the complexities of commercial radio's response to modernity during the interwar years. The quintessence of modernity, radio had the capacity to deny its own status as a mass medium. Griffen-Foley demonstrates how early Australian radio blended the rhetoric of modernity with compensatory varieties of 'personal' contact, with 'uncles', 'aunts' and 'friends'—people 'just like you and me'—carefully ministering to their audiences. Reaching out to their listeners by considering their problems, leading them in community singing concerts, establishing social and charitable clubs and appearing at special events, radio stations transformed themselves self-consciously from a modern technological mass medium into a unifying, intimate and highly personalised family companion.

Brian Yecies completes the volume by analysing local Australian responses to the coming of sound to the cinema in the late 1920s and early 1930s. He shows how the global transition to sound was more local than previously thought, challenging conventional assumptions about global and local business interests in the cinema industry. Tracing how the introduction of sound was influenced and shaped by a diverse group of Australians, he demonstrates how they were caught between the promise of global modernity, collusion and the threat to national identity. Yecies shows how local firms did innovate, adopt, thrive and survive—not by defeating the Americans but by assimilating them, by making them Australian rather than shaping Australian cinema in a solely American way. Whether or not they realised it, Yecies argues, they were agents of modernity, mediating complex social, cultural and technological changes in Australia.

ENDNOTES

[1] Corbin, Alain 1998, *Village Bells: Sound and Meaning in the 19th-Century French Countryside*, trans. by Martin Thom, Columbia University Press, New York, pp. xi–xii; Connor, Steven 1997, 'The Modern Auditory I', in R. Porter (ed.), *Rewriting the Self: Histories from the Renaissance to the Present*, Routledge, London and New York, pp. 203–23, esp. p. 219. For the primacy of sight see Jay, Martin 1993, *Downcast Eyes: The Denigration of Vision in Twentieth-Century French Thought*, University of California Press, Berkeley; Levin, Michael (ed.) 1993, *Modernity and the Hegemony of Vision*, University of California Press, Berkeley; and Ong, Walter J. 1991, 'The Shifting Sensorium', in D. Howes (ed.), *The Varieties of Sensory Experience: A Sourcebook in the Anthropology of the Senses*, University of Toronto Press, Toronto, pp. 47–60. For the new anthropology and history of the senses, see Feld, Steve 1982, *Sound and Sentiment: Birds, Weeping, Poetics, and Song in Kaluli Expression*, University of Pennsylvania Press, Philadelphia and Feld, Steve with Keith H. Basso (eds) 1996, *Senses of Place*, School of American Research Press, Santa Fe, NM; Classen, Constance 1993, *Worlds of Sense: Exploring the Senses in History and Across Cultures*, Routledge, London; Kahn, Douglas 1999, *Noise, Water, Meat: A History of Sound in the Arts*, MIT Press, Cambridge and Kahn, Douglas 2000, 'Sound awake', *Australian Review of Books*, July, pp. 21–2; Douglas, Susan 1999, *Listening In: Radio and the American Imagination*, Times Books, New York; Smith, Mark M. 2001, *Listening to Nineteenth-Century America*, Chapel Hill and Smith, Mark M. 2003, 'Making sense of social history', *Journal of Social History*, vol. 37 no. 1, pp. 165–86; Thompson, Emily 2002, *The Soundscape of Modernity: Architectural Acoustics and the Culture of Listening in America, 1900–1933*, MIT Press Cambridge. Australian historians Shane White and Graham White recently made a distinguished contribution to this literature with *The Sounds of Slavery*, The Beacon Press, Boston (2005).

[2] Atkinson, Alan 2002, *The Commonwealth of Speech: An Argument about Australia's Past, Present and Future*, Australian Scholarly Press, Melbourne, pp. xii, xvii; Damousi, Joy 2005, *Freud in the Antipodes: A Cultural History of Psychoanalysis in Australia*, University of New South Wales Press, Sydney, p. 4.

1. A 'Roaring Decade': Listening to the Australian gold-fields

Diane Collins

Unlike lions or trade winds, decades do not, in general, roar. An obvious exception is the 1920s. The roar of the Twenties came from new technologies such as radio and talking pictures, new art forms such as jazz, a more strident and intense urbanism and new personal freedoms. Sound was listened to as inescapable and affirmative, the language of machines, experiment, consumption, pleasure. Except for such rare cases, it is not sufficiently appreciated how our experience of sound influences our interpretation of time. The notion of the 'Roaring Twenties' is, to be sure, a metaphor, but historians and other social analysts have not been sensitive enough to time's aural dimension. Certain periods have an intensity of change linked to popular experience that is aural as much or more than it is visual. We must therefore explore 'aural time': the conceptualisation and periodising of history according not merely to visual and material changes reflected in documents and artefacts but to the experience of sound.

The deeper exploration of aural time casts light on the nature of modernity. Cultural critics have taken the intrusion of industrial sound as part of the distinctive contribution of the 1920s and of the movement of modernity.[1] Historians of Australia can provide a different perspective. What if these ideas of the link between sound as a revolutionary signifier of modernity were applied to key changes in Australian society?

Certainly, the experience of sound in pre-industrial society was different from modern technological civilisation. Much attention has been devoted to the sound-scapes of traditional European society, while the models of modernity assumed for Europe do not necessarily work for Australia.[2] There were, and are, multiple modernities and the coming of the new was a protracted, uneven and problematic process. In some senses, at least, the 1920s relived much earlier historical experiences of sound as rupture. Seventy years earlier, for example, in Australia, another decade also roared and, as in the 1920s, this roar was linked to new material environments and new behaviours.

This discussion listens to a range of 1850s gold-field acoustics: industrial and social noise, silence, body sound-scapes and music. I am interested in the general relationships of gold-field sounds to the emergence of modernities, the contestation over the meaning of gold-field sound-scapes and the way some sounds have been silenced or recalled as memory. The processes of absorption increasingly involved—in recollections or immediate experiences of the gold-fields—a refusal to hear what Malcolm Gillies has called 'alternative Australias'.[3] Who recalls, for example, that one of the most obvious additions to the sound-scape of colonial Australia in the gold-rush period was the ubiquitous sound of gun-fire? In this regard, I have sought to listen particularly to the silencing of guns in the ways that the gold-fields were portrayed and remembered.

Most problematic were the messages drawn by contemporaries from the gold-fields' tumultuous auditory world. With up to 30,000 diggers on some alluvial gold-fields, large diggings were heard long before they were seen or felt. Nearing the Sofala strike, Godfrey Mundy's companion suddenly cried, 'Stop and listen.' Mundy heard what he imagined was the 'rushing of some mighty cataract'. The extraordinary noise came from stones grating on the metal sifters of 500 cradles, box-like devices designed to separate gold from the surrounding sand and gravel. So 'unforgettable' was the 'uniform and ceaseless crash' that he marvelled that it came from 'the agency of human beings, not one of whom was visible'.[4] On reaching the diggings, gold-seekers heard other discordances: the clanging of picks and shovels in shafts, the wheeling of barrows, the slurp of water in thousands of gold pans. But while the diggings were acknowledged universally as a distinctive acoustic spectacle, the gold-field positioned listeners in complex and contradictory ways. Even the diggings' industrial sound-scape produced, for example, the same deep ambivalences of the larger gold-rush period. For some listeners, the sonic intensity spurred a positive revaluation of the natural sound-scapes of Australia, a desire to escape not from, but into, the quiet of the natural world. Ellen Clacy and her brother eagerly left the gold-field one Sunday morning intent on wandering into spots 'where the sound of the pick and shovel, or the noise of human traffic, had never penetrated'. They found the scene 'harmonious, majestic, and serene'. The mighty forests 'hushed in a sombre and awful silence' and the 'soft still air' led Clacy 'from the contemplation of nature to worship nature's God'.[5] As he left the diggings, Thomas McCombie heard the 'sounds of riot and debauchery' die gradually away and he entered 'with pleasure, the open glades of the forest, where not a sound was to be heard'.[6] For some, even the 'sharp wiry notes' of some of Australia's more dissonant birds sounded 'far more agreeable than the barking of dogs and the swearing of diggers'.[7] While hardly congruent with modernity's rejection of naturalism, this change in environmental sensibility was an important emotional and aesthetic progression in the inscription of value on Australian landscapes.

Acoustically, modernity gained definition by the continual creation of new silences as well as new sounds. In the 1920s, for example, Paris was depicted popularly and simultaneously as an icon of the roaring decade and as 'a quiet and uneventful village', the two discourses so integral to each other as to collapse the dichotomy between avant and rearguard, modernity and reactionary modernism.[8] In the 1850s Victorian gold-fields, we can listen to an earlier working out of this interplay. Because of the tight knit between acoustic and social calm, the din of gold did promote a backwards listening, an idealising of the imagined quiet and order of pre-modern pastoral sound-scapes. But there was no retrospectivity in the way gold-fields' cacophony retuned the colonial ear to hear new meaning in one of the most ubiquitous and troubling aspects of colonial Australia: the gnawing silence of the land.[9] Explorers had heard this silence as void, a corrosive muteness that mocked the psyche's search for meaning. But this was a silence of the mind and, when heard in contrast with the bedlam of the gold-field, men and women found in this apparent noiselessness a source of uplift. In silence, they now heard not emptiness but moral, aesthetic and even spiritual affirmation.

If the once corrosive silence of the bush was transformed, for some, into a sound-scape of refuge and uplift, it is clear that not everyone listened to the noise of gold-seeking as nuisance or discordance. For many, mining was an aesthetically satisfying acoustic. The 'sound most pleasing to the digger's ear was the rattle of the cradle as it rocks'[10] 'It's great to listen to the miner's cradle, rattle, rattle from side to side,' remarked one adventurous female digger, '...and with anticipation wonder if there is gold at the bottom of it all.'[11] An immediate explanation was that gold-seeking represented the satisfying sounds of human enterprise. But the frenetic activity at the auction rooms in Ballarat evoked images of 'dreadful hubbub'[12] and there was not much about gold-seeking that resembled the idealised acoustics of manufacture, the 'quiet, plodding' spirit of 'regularly rewarded industry'.[13] Rather, the relentless acoustic of alluvial mining was heard as opportunity, as evidence that the economic and moral trajectory of the nineteenth-century economy could be release. The miner was an independent labourer who gained rewards through hard but mostly honest toil and owned the wealth he produced. So every rattle, every shake, chorused the possibility of individual transformation, of an escape from destiny to freedom.

For many visitors to the gold-fields, the most audible noise was not, however, the deafening crash of the miners' cradles but the social sound-scape. The meaning of this sound-scape was contested especially with conservative listeners likely to hear only a moral din that pointed to deepening social degeneracies: alcoholism, collapsing domesticity, godlessness and avarice. But whether listeners were censorious or approving, the gold-fields were heard as a modernity in which the auditory environment signalled less a technological transformation than a re-engineering of the sound-scapes of the self. As economic independence translated into social independence, and migration loosened customary bonds, linguistic and bodily behaviours were redrawn. Like the search for gold, this independence required announcement, a public claim. The much commented on prevalence of blunt greetings, swearing, rude jests, plain speech, hearty laughter and loud, direct, unapologetic conversation on the gold-fields record the extent to which the digger staked out a newly perceived autonomy by democratising the realm of sound.[14]

A number of historians have paid attention to the politics of language in early colonial Australia. The demographic and economic disorder of the 1850s gold-fields created the circumstances for an intense experiment with sound, propriety and power. Not only was a once essentially transgressive use of language and acoustic gesture increasingly assimilated into individual behaviour, the instability of aural relationships contributed to the 'problem of the diggings'. The infamous licence hunts revolved around a succession of acoustic gestures in which miners and police played out notions of law and legitimate resistance in a vivid theatre of sound.[15] Within the diggings, the law enforcers, in particular, seem to have envisaged their duties as aural combat. 'To perform their duty quietly,' one of the gold commissioners subsequently remarked of the diggings police, 'without bouncing, bullying and swearing at everyone; to ask a man quietly whether he had a licence was out of the question, it must be accompanied by some low life expression.'[16] The lack of respect for police on the diggings and the problems this posed were tied intimately to the law enforcer's own disruptive sound-scapes. In Victoria, conflict was as surely rooted in the contest over aurally constructed systems of deference and hierarchy as in the government's administration of the diggings.

In the preference for clear, functional and direct sound, the diggings' acoustic etiquette was characterised by qualities integral to the modernising of the social landscape and the self as sound. But while the 1920s increasingly constructed choice, consumption and desire through auditory cultures, the 1850s meshed sound, liberty and subjectivity in a more elemental relationship. From the miner's point of view, acoustic freedom was heard as part of the Enlightenment's rights discourse and din was inherent in the transition to the broad spectrum of democratic forms. In this, the roar of the gold-fields resembled the auditory conflicts of the revolutionary era as much as the dissolutions of the Twenties. For a time in the 1760s and 1770s, American colonists, for example, similarly demanded that 'the elocution of gentlemen was to be replaced...by the genuineness and candour of common speech'.[17] For a time, at least, this assertion of the 'plain language of free men'[18] was fundamental to the search for a revolutionary sensory model that could sustain the claims for republican citizenship.

In different historical circumstances, gold-field diggers also sought to define the acoustic personality and aural culture required of a free economic, as much as political, man. The importance attached to the acoustic assertion of a more aggressive and self-conscious citizenship is measured not only by the diggers' words and body language but by a continual resort to aural props that gave the digger's 'voice' additional authority and menace. As much as through the adoption of more direct forms of speech and auditory gesture, it was these strategies for extending the physical limits of voice that amplified gold-field experiments with the emancipation of the self.

Every digging, for example, was marked by the prevalence of 'hosts of mongrel dogs of the most noisy and snappish breed'. Memories of incessant dog barking survive in all reminiscences of 1850s gold-seekers and the impact of this canine artillery was clear. The uncompromising blend of viciousness and volume not only made the dogs useful security guards as well as companions but, in a world where the barriers between the public and private were flimsy, dog barks were more effectual than tent walls in defining and enlarging the sense of private space. A dog's bark was a sonic barrier, the breach of which was trespass. The ownership of these dogs was also part of the diggers' usurpation of the trappings of gentility since, for centuries, dog-keeping formed part of aristocratic culture.[19] While visitors often found the yapping intensely aggravating, from the diggers' point of view the barking of dogs was a noise resonant with ambivalence, an admission of insecurity and loneliness and a belligerent enactment of new autonomy.

The most crucial item in a miner's sonic ammunition was not, however, barking dogs but guns. Although every gold-field resounded with the sound of gun-fire, posterity's image of the gold digger in a cabbage-tree hat, intently bent over a gold pan, blots out the centrality of firearms. Antoine Fauchery noted that every digger was 'armed to the teeth'.[20] John Sherer advised his Old World readers that 'pistols—deadly revolvers—ought to be the companion of all emigrant diggers'.[21] Like many gold-seekers, Arthur Polehampton regretted that he had not brought several dozen ordinary pistols with him since he could have sold them for a substantial profit. Many new arrivals were, he observed, armed to such an extent 'that they might very easily have been mistaken for bushrangers'.[22] Though many firearms were expensive and in short supply, the less-efficient 'miner's pistol' could be bought for 5/6d. Imported from Belgium, this single-fire gun could be loaded with ball, pebble, shot or stones. Many miners carried such pistols stuck in their belts though Americans were more likely to sport Colt 45s.[23] Since theft was common and crime widespread, firearms offered the digger protection and defence from the ubiquitous fear of robbery.

The prevalence of guns explains one of the most characteristic sounds of Australia's gold-field: the discharge of guns every evening. This ritual fusillade began with occasional shots, growing in intensity until the sound more resembled flanking fire. The 'perpetual thunder' lasted well into the night until, eventually, the shots became more widely spaced, finally ceasing, as it began, with occasional discharges.[24] The practice ('as if Royal Princes were being incessantly born')[25] also gave the diggings a decidedly military air. One observer noted: '[A] stranger coming to the mines at night would fancy that he was approaching a battle-field. The numerous groups of armed men congregated around their camp-fires, and the incessant rattle of discharging arms for hours together, form no bad emblem of a war-scene.'[26] Fauchery considered that every evening during this 'veritable insurrection', the diggers fired more rifle and pistol shots than were fired in the European revolutions of July 1830.[27]

There was a utilitarian explanation of this practice. Guns were fired and reloaded in order to keep them clean and functioning. William Howitt, however, thought that, except in rainy weather, the powder was unlikely to become damp and found just as convincing the idea that the diggers, like children, were simply 'immensely delighted with the noise of gun-powder'. The 'abominable' din was, he believed, simply 'for the sake of noise', a view bolstered with claims that, in the Old World, very few of the diggers would have ever handled a gun. Here, however, they all had them and 'are out, and firing at everything they see'.[28]

Howitt was not alone in this assessment. Seweryn Korzelinski likewise thought the firing away of old cartridge and reloading a 'superfluous activity' if the arms were kept dry. In an environment where gold-seekers were drawn from the four corners of the globe, shooting seemed more about unambiguously announcing to robbers that there was a gun in the tent or, alternatively, of manufacturing Dutch courage, self-assurance from the need to worry about an assault.[29] But the practice made Miska Hauser uncomfortable as—like Howitt—he considered many diggers were 'dilettantes' with no more experience with firearms than he had with changing babies. Hauser also regretted the diggers' propensity to fire randomly in all directions, including into miners' tents—a practice that had resulted in sleeping miners being killed by stray bullets.

An enchantment with gun-fire pervaded the Australian gold-fields of the 1850s. Men traditionally precluded from the ownership of firearms exulted in their possession of an object once so firmly associated with aristocratic culture. Many diggers' guns hardly rated as reliable armaments and this also encouraged the theatrical use of firearms. Guns became most valuable as sound, as a ritualised auditory demonstration of the way in which gold renegotiated deference and independence. Amid the linguistic and cultural diversity of the gold-fields, gun-fire functioned as a shared language, serving as a militarisation of private voice and also as evidence that the highly individualised voices of the gold-field could coalesce into a community of independent speakers. But gun-fire was more than a rival expression of authority, a refusal to accept subordination. Like other acoustic gestures, the theatre of gun-fire, in privatising the display of power, underscored the gold-field as an experimental democracy of sound. It is no wonder that visitors explicitly compared the diggings with such emblematic sites of sonic disorder as Pandemonium, Hades, Bedlam and the Greenwich and Donnybrook Fairs.

In the shifting, socially diverse but visually homogenous world of the gold-fields, modernity's sensory hierarchy was disrupted. In communicating conceptions of self and society, community and difference, sight surrendered primacy to sound. Yet the sound-scape was tumultuous. Theorists sometimes claim that, unlike the eye, the ear is an indiscriminate receptor. But to hear sound as social knowledge, the gold-field's resident had to engage in a good deal of selective listening. A character in the gold-rush novel *Clara Morison* observed, for example, that voice 'is the great means of recognition all over the diggings'.[30] While this sound became the most immediate source of information about a man or woman's background, sound—organised as music—also became crucial in the search for meaning. On one level, music functioned as community. The much remarked on nightly camp-fire songs provided release from the patent monotony and frustration of mining life and, like gun-fire, gained currency as a democratic acoustic. Because diggers often congregated in national groups, gold-field songs and instrumental music were also listened to as a way of narrating difference. Within a culture marked by fragmentation and dislocation, music was heard to speak of origins and the result was a sonic delineation of identity in which listening was sometimes tied to imaginings of race but, more often, to a construction of identity in which the tribal and national were fused. Gold-field music was thus heard as a diasporic sound in which nationality was frequently more audible than tune. But in this listening, the dissonance of nations could also be harmonised to represent less difference or conflict than the auralities of a possible future cosmopolitanism.

The capacity of music to organise subjectivity[31] made music an acute barometer of the tension between abandonment and belonging inherent in many nineteenth-century and colonial explorations of identity. Returning to his tent in Kangaroo Gully, W. Craig listened to a song that touched him deeply. A party of four miners sat around their camp-fire, one of their number singing the latest popular ballad, *Ben Bolt of the Salt Sea Wave*. The miner sang feelingly in a deep, rich voice that reverberated through the gullies of Bendigo:

> There's a change in the things that we loved, Ben Bolt;
> There's a change from the old to the new.

Fifty years later, Craig remembered the moment:

> Of all the crowd that assembled then
> There's only you, mate, and I.[32]

In this song, the wider meaning of gold-field independence sounded most acutely. Old securities and relationships fell away suddenly and in their place was the aloneness of each digger. Of course, not all the music on the gold-fields came from miners in their tents. Against the nostalgic assertions of lost folk communities, commercial musicians such as Charles Thatcher commodified new colonial identities through songs that fused gold-seeking and colonial experience, sound and place. Miners could, however, be an unsympathetic audience. When the violinist Miska Hauser, whose violinist father was a friend of Beethoven, visited Australia, his concerts at Ballarat produced noisy disapproval whenever he departed from the anticipated repertoire: *Carnival*, *My Little Bird* and *The Song of Tahiti*.[33] For all the talk of digger independence, there also operated a vicious exclusionary ethic that demanded conformity to certain aural (and visual) practices that gave identity to miners as a group.

Sounds provided not only the architecture of identity on every diggings, but much of the substance of gold as memory. Poems, short stories, a television series and the recent musical *Eureka!* recapture the 1850s as a decade in which sound was essence (the most evocative aspect of the TV series was the theme music). Typical of early acoustic memorialisation is Henry Lawson's short story *Golden Gully*, published in 1887. More than the gold had gone from the gully. 'The "predominant note" of the scene' was a 'painful sense of listening, that never seemed to lose its tension, a listening as though for the sounds of digger life, sounds that had gone and left a void that was intensified by the signs of a former presence'.[34] In Lawson's recollection of the gold-fields, the visible evidence of the rush remained but could not be understood without the lost sound-scape. Without sound, neither memory nor the viewer could truly perceive.

Since sound was more transitory than image it seems likely that, in Australia and California, the 'very early appearance of nostalgia for the days of gold'[35] was tied specifically to recognition of aural change so extensive that it ranged from environmental sound to language. When Mark Twain visited Ballarat 45 years after the great rushes, he noted that a post-gold process of linguistic cleansing had produced 'Ballarat English', a speech free from Americanisms, vulgarisms or any kind of emphasis.[36] Like the 1920s, the gold-fields roared louder and more vividly because the years immediately following were listened to as a quieter and leaner acoustic time. As an affirmative site of acoustic memory, the gold-fields became notable for symbolising not the paradigmatic yearning for a vanished rural quiet but for an alternative rather than counter urbanism. This was a world of dense and constant sound in which the incompatibility of men and machine dissolved, listening realigned with possibility and the solitary introspective voice existed within a fellowship of sound, a multiplicity of voices, yet each distinct, each capable of being heard.

It is a further irony that when historians re-listened to the gold-fields they remained silent about a range of sounds. Among historians, the gold-fields were constructed through numerous acoustic omissions in relation to alternative sound-scapes— including race, environment and gender—but one of the most provocative examples of subsequent historical deafness is, arguably, the silencing of gun-fire. In historical writing, this sound was virtually forgotten. For example, 1963 saw the publication of Geoffrey Serle's *The Golden Age*, which was and is the major work on Australia's gold-rush period. The book contains no index entry for guns and only fleeting textual references. The same year saw a new edition of the *Australian Encyclopedia*. This impressively comprehensive publication contains no entry on guns or firearms. *Gold Seeking*, David Goodman's revisionist 1994 study, pays even less attention to this subject, while the 2001 work *Gold: Forgotten Histories and Lost Objects of Australia* [37] sustains historical amnesia on the topic. An unashamed gun lobbyist has produced the only explicit if essentially amateur work on gold and firearms.[38]

Nor does the subsequent attention to Eureka contradict this general acoustic amnesia. Despite later songs and poems, it is telling that this rare instance of armed and fatal resistance to the colonial state has been assimilated into memory so strongly as sight not sound, as the image of the Eureka flag. To the extent that Eureka's gun-fire remains part of historians' accounts it is because of its exceptional status, its association with bloodshed. The deeper sounds of Eureka are, however, related less to the romance of nation than to Australia's evolution as an English polity. Unlike the Americans and some Europeans, the English were a disarmed people.[39] Australian history was similarly heard as free of any acoustic glorification of an armed citizenry able to act outside the State. In this highly strategic listening, gun-fire, in particular, was later remembered largely in moments when it could be sanctified, as with the distant shell-fire of World War I, as national sound. Though bursts of gun-fire could also be heard among the criminally inclined, gold-rush evidence of a larger acoustic narrative relating guns, an incipient individual rights culture and violence to the emergence of the nation was virtually extinguished.

Inevitably, the relationship between gold-rush sound and modernity is imprecise. Resonances of future technological sound-scapes were stronger when deep shaft-mining came to predominate, bringing the throb of steam engines to the gold-fields and mine whistles to mark shift changes. Alluvial mining was more acoustically ambiguous since the tools were relatively primitive, often dependent on muscle power and, in some cases, still tied to the cycle of human breath. And, after work had ceased, one of the most dominant sounds on an alluvial gold-field was, as in pre-industrial times, the human voice. It was the silencing of these voices, of a sense of acoustic community, that made many memories of the golden era so melancholy. Although modernity's separation of work and play was also heard on the diggings, the sounds of play remained linked to a time when amusement was more out-of-doors, communal and self-made. Diggers created most of their entertainment and music-making remained central to entertainment. But, in opposition to the piano in the parlour—the iconic symbol of nineteenth-century domestic music—the gold-field was a world in which flutes and accordions predominated. These were instruments of less settled and technologically simpler societies and, in many ways, the sound-scape of the diggings clearly hark back to nomadism. Yet, this wandering now took place in a genuinely global context and the acoustic tensions between individualism, community and state, fragment and core, ephemerality and permanence were heard on every diggings. In this, and other ways, the gold-fields' sound-scapes were tied most closely to the emotional history of modernity.

The 1920s roared because of the decade's relationship to the immediate past as well as to coming sounds, especially those of the Great Depression. Instead of the gun-fire of World War I—the sound of death, destruction and human sacrifice—came the sounds of technologies of pleasure. The 1850s roar gained resonance from contrast with the harsh and disciplining sound regimes of the preceding convict era. But the roar of the 1850s came also from a heady experiment with the senses. In diverse ways, the gold-fields' private and public sound-scapes became agents for an exploration of modernity's emerging 'rights' culture and also its slowly deepening preoccupation with subjectivity. Acoustically, the 1850s did encourage a new hearing of the silences of Australia but, on the gold-fields, human and manufactured noise was also, for a short while, not only celebrated but aestheticised as freedom. Then, as in the infant American republic, there was a return to moderation and sedateness as conservatives reasserted the virtues of tranquillity. Peter Hoffer has drawn attention to the ways in which, in the United States, social sounds were homogenised and flattened and discourse desensitised.[40] In post-1850s gold-rush Australia, incipient colonial democracies no longer clanked with convict chains but acoustic fetters now bound more tightly the impulse for social experiment and especially the embrace of individualism, the privileging of personal liberty above collective stability as social good.[41]

This chapter is a small part of a larger study of sound in nineteenth and early twentieth-century Australia. Many further questions—about the relationship between gender, gold and sound, for example—could be asked. But it is clear that, in gold's roaring decade, sound and history stood in a close relationship. In this decade, the meaning of 'modern' time was both ruptured and intensified by sound. Some commentators have argued against too glib an equation between modernity and the uninterrupted privileging of the visual. This argument is drawn from technological evidence. The emergence and development of loud artillery and a myriad inventions such as the telephone, gramophone, subway train and automobile are offered as proof of the primacy of the aural in some modern periods, such as the years between 1870 and 1914.[42] While there is much to persuade in this analysis, such an approach neglects the impact of other types of social change. In gold-rush Australia, mineral discovery proved as powerful as loudspeakers or electric motors in challenging and advancing modernity and its imagined relationship to the senses.

ENDNOTES

[1] See, for example, Thompson, Emily 2002, *The Soundscape of Modernity: Architectural Acoustics and the Culture of Listening in America 1900–1933*, MIT Press, Cambridge, Mass., and London; Erlmann, Veit (ed.) 2004, *Hearing Cultures: Essays on Sound, Listening and Modernity*, Berg, Oxford and New York; Morton, David 2000, *Off the Record: The Technology and Culture of Sound Recording in America*, Rutgers University Press, New Brunswick, NJ, and London; Golan, Romy 1995, *Modernity and Nostalgia: art and politics in France between the wars*, Yale University Press, New Haven and London.

[2] On Euro–American sound-scapes, see Corbin, Alain 1998, *Village Bells: Sound and Meaning in the Nineteenth-Century French Countryside*, Columbia University Press, New York; Smith, Bruce 1999, *The Acoustic World of Early Modern England: Attending to the O-factor*, University of Chicago Press, Chicago; Schmidt, Eric Leigh 2000, *Hearing Things: Religion, Illusion and the American Enlightenment*, Harvard University Press, Cambridge, Mass.; Smith, Mark M. 2001, *Listening to Nineteenth- Century America*, University of North Carolina Press, Chapel Hill and London; Rath, Richard Cullen 2003, *How Early America Sounded*, Cornell University Press, Ithaca and London; Hoffer, Peter Charles 2003, *Sensory Worlds in Early America*, John Hopkins University Press, Baltimore and London; White, Shane and White, Graham 2005, *The Sounds of Slavery: discovering African American history through songs, sermons and speech*, Beacon Press, Boston.

[3] Gillies, Malcolm 2001, 'Alternative Australias: Fates and Fortunes', *Proceedings of the Australian Academy of the Humanities*, Canberra, pp. 134–50, at p. 142.

[4] Mundy, Godfrey Charles 1853, *Our Antipodes: or residence and rambles in the Australasian Colonies*, 3 vols, Vol. 3, Richard Bentley, London, pp. 369–70.

[5] Clacy, Mrs Charles [Ellen] 1963, *A Lady's Visit to the Gold Diggings of Australia in 1852–53*, Landsdowne Press, Melbourne, first published 1853, p. 72.

[6] McCombie, Thomas 1861, *Australian Sketches*, W. Johnson, London, p. 20.

[7] Howitt, William 1972, *Land Labour and Gold*, Lowden, Kilmore, first published 1855, pp. 269–70.

[8] Golan, *Modernity and Nostalgia*.

[9] See, for example, Collins, Diane 2006, 'Acoustic journeys: exploration and the search for an aural history of Australia', *Australian Historical Studies*, Vol. 37, pp. 1–17.

[10] Sherer, John (ed.) 1973, *The Gold Finder of Australia*, Penguin, Harmondsworth, first published 1853, pp. 39, 35.

[11] Tyler, Mary Ann 1985, *The Adventurous Memoirs of a Gold Diggeress, 1841–1904*, published privately by Kate Gibbs, p. 38.

[12] Hauser, Miska 1988, *Letters from Australia 1854–1858*, edited by Colin Roderick and Hugh Anderson, Red Rooster Press, Maryborough (Vic.), p. 61.

[13] Kelly, William 1977, *Life in Victoria*, Vol. 1, Lowden, Kilmore, first published 1859, p. 38.

[14] Sherer, *The Gold Finder of Australia*, p. 75.

[15] See, for example, *Among the Diggers* 1999, Ballarat Heritage Service, Ballarat, p. 9.

[16] Read, Charles Rusdon in Geoffrey Serle 1963, *The Golden Age: A History of the Colony of Victoria 1851–1861*, Melbourne University Press, Melbourne, p. 99.

[17] Hoffer, *Sensory Worlds in Early America*, p. 242.

[18] John Dickinson, 1781, quoted in ibid., p. 224.

[19] For a discussion of dogs and European culture in the nineteenth century, see: Kete, Kathleen 1994, *The Beast in the Boudoir: Pet Keeping in Nineteenth-Century Paris*, University of California Press, Berkeley; Ritvo, Harriet 1987, *The Animal Estate: The English and Other Creatures in the Victorian Age*, Harvard University Press, Cambridge, Mass.

[20] Fauchery, Antoine 1965, *Letters from a Miner in Australia*, Georgian House, Melbourne, first published 1857, p. 53.

[21] Sherer, *The Gold Finder of Australia*, p. 173.

[22] Penzig, Edgar 1993, *Guns and Gold: Stories, Artifacts and Crime on the Australian Diggings 1850–1900*, Tranter Enterprises, Katoomba, p. 45.

[23] Ibid., p. 13.

[24] Hauser, *Letters from Australia 1854–1858*, pp. 64–6; Howitt, *Land Labour and Gold*, p. 188.

[25] Burchett, F. quoted in Serle, *The Golden Age*, p. 79.

[26] Sherer, *The Gold Finder of Australia*, p. 182.

[27] Fauchery, *Letters from a Miner in Australia*, p. 53.

[28] Howitt, *Land Labour and Gold*, pp. 188–9.

[29] Korzelinksi, Seweryn 1979, *Memoirs of Gold-Digging in Australia*, edited by Stanley Robe, University of Queensland Press, St Lucia, p. 64.

[30] Spence, Catherine Helen 1971, *Clara Morison*, Rigby, Adelaide, first published 1854, p. 234.

[31] Steinberg, Michael P. 2004, *Listening to Reason: Culture, Subjectivity and Nineteenth Century Music*, Princeton University Press, Princeton and Oxford.

[32] Craig, W. 1903, *My Adventures on the Australian Goldfields*, Cassell, London, p. 231.

[33] Hauser, *Letters from Australia 1854–1858*, p. 67.

[34] Lawson, Henry 1984, 'His Father's Mate', in Leonard Cronin (ed.), *A Camp-Fire Yarn: Henry Lawson Complete Works, 1885–1900*, Lansdowne Press, Sydney, pp. 56–7.

[35] Goodman, David 1994, *Gold Seeking in Victoria and California in the 1850s*, Allen & Unwin, Sydney, p. 2.

[36] Twain, Mark 1973, *Mark Twain in Australia and New Zealand*, Penguin, Harmondsworth, first published 1897, pp. 237–8.

[37] McCalman, Iain et al. (eds) 2001, *Gold: Forgotten Histories and Lost Objects of Australia*, Cambridge University Press, Cambridge, UK, and New York.

[38] Penzig, *Guns and Gold*.

[39] Gilmour, Ian 1992, *Riot, Risings and Revolution: Governance and Violence in Eighteenth-Century England*, Hutchinson, London, p. 19.

[40] Hoffer, *Sensory Worlds in Early America*, pp. 243–4.

[41] For a discussion of the re-quietening of Australia see Goodman, *Gold Seeking in Victoria and California in the 1850s*.

[42] See, for example, Schwartz, Hillel 2003, 'The Indefensible Ear: A History', in Michael Bull and Les Black (eds), *The Auditory Culture Reader*, Berg, Oxford and New York, pp. 487–501.

2. A Complex Kind of Training: Cities, technologies and sound in jazz-age Europe

James Donald

Writing in the 1920s, Robert Musil opens his novel *The Man Without Qualities* by evoking Vienna in the final days before the Great War changed everything. His visual imagery recalls contemporary experiments in abstract and rhythmic film-making by Walther Ruttmann, Hans Richter or Viking Eggling. Blocks of light and shade are cross-cut by lines in motion—speeding automobiles in Musil's case—while the movement of pedestrians negotiating their way through the city's streets creates more fluid, fractal patterns. The way Musil 'hears' the city, however, is more ambivalent and less assertively modernist. His 'wiry texture' of sound shares the jagged angularity and dynamism of the visual images, but the way in which he then anchors this particular soundscape to Vienna seems to hark back to an earlier, more traditional aesthetic that assumed sound to be characteristic of, and specific to, place.[1]

> It was a fine day in August 1913.
>
> Automobiles shot out of deep, narrow streets into the shallows of bright squares. Dark clusters of pedestrians formed cloudlike strings. Where more powerful lines of speed cut across their casual haste they clotted up, then trickled on faster and, after a few oscillations, resumed their steady rhythm. Hundreds of noises wove themselves into a wiry texture of sound with barbs protruding here and there, smart edges running along it and subsiding again, with clear notes splintering off and dissipating. By this noise alone, whose special quality cannot be captured in words, a man returning after years of absence would have been able to tell with his eyes shut that he was back in the Imperial Capital and Royal City of Vienna. Opening his eyes, he would know the place by the rhythm of movement in the streets long before he caught any characteristic detail. It would not matter even if he only imagined he could do this.[2]

Had this imagined exile returned 10 years later—when Musil was writing—would the sound of the city still have identified it so immediately and unchangingly as Vienna and nowhere else? And would our exile, with eyes shut or wide open, have listened to the city in the same way? It is doubtful. He would now have found himself in Red Vienna, struggling to recreate itself in the straitened aftermath of war and revolution, rather than in Imperial and Royal

Vienna in its final spasms of self-deluding pomp. The purr and jangle of ostentatious wealth and the brassy *folies de grandeur* of autocratic power would to a large degree have been silenced. More striking would have been the noisy business of building a liveable social democratic city, and the sounds of a city opening up to more demotic, international and commercial forms of culture and communication. Vienna's particular political make-up, its demographic changes, the speed of its economic recovery, the particular emphasis on building workers' housing and constructing a social infrastructure, the density of public and private transport, the new types of leisure and entertainment venues to be offered to the public and the emergence of new media technologies—all these factors, and others, would have combined to produce a distinctive mix that would have made the city still sound different from Berlin, Paris or London. Equally, however, new sounds associated with social changes were increasingly common to all these cities, and so their soundscapes were inevitably becoming more and more similar. Perhaps as a result, the idea of sound as the acoustic genius of a place was giving way increasingly to the idea of noise as an intrusive and wiry-textured aspect of the urban environment that needed to be measured, managed and controlled.

This new, more modernist style of listening becomes dominant as Musil switches the emphasis of his narration from the uniqueness of Vienna's historical soundscape to its modern typicality:

> So let us not place any particular value on the city's name. Like all big cities it was made up of irregularity, change, forward spurts, failures to keep step, collisions of objects and interests, punctuated by unfathomable silences; made up of pathways and untrodden ways, of one great rhythmic beat as well as the chronic discord and mutual displacement of its contending rhythms. All in all, it was like a boiling bubble inside a pot made of the durable stuff of buildings, laws, regulations, and historical traditions.[3]

The sociological imagination behind these metaphors is still that of Georg Simmel, whose 1903 essay, 'The Metropolis and Mental Life', had not only prefigured Musil's theme of the loss of 'qualities' (or the individual, subjective characteristics Simmel termed 'personality'), it had offered an explanation for the phenomenon. 'The individual,' wrote Simmel, 'has become a mere cog in an enormous organisation of things and powers which tear from his hands all progress, spirituality, and value in order to transform them from their subjective form into the form of a purely objective life.' What he calls the 'overwhelming fullness of crystallised and impersonal spirit' is identical to Musil's 'durable stuff of buildings, laws, regulations, and historical traditions'.[4] The difference between the two accounts lies in the novelist's ability to convey what the implosion of 'personality' felt like by presenting an almost synaesthetic conflation of the social with the sensory—with vision and movement, but also with hearing and sound. Modernity in Vienna was experienced as cacophony and syncopation: 'the chronic discord and mutual displacement of its contending rhythms.'

Simmel's pupil Walter Benjamin also tried to capture the somatic or sensory level at which people felt and adapted to the rhythms of a changing world. In *Charles Baudelaire: A Lyric Poet in the Era of High Capitalism*, he distinguishes between Poe's nineteenth-century 'Man of the Crowd' and the contemporary pedestrian whose habits of looking, seeing and movement have been changed by the development of motor traffic and even by the traffic signals that first appeared in Berlin's Potsdamer Platz in 1926.

> Moving through [the traffic of a big city] involves the individual in a series of shocks and collisions. At dangerous crossings, nervous impulses flow through him in rapid succession, like the energy from a battery. Baudelaire speaks of a man who plunges into the crowd as a reservoir of electric energy. Circumscribing the experience of the shock, he calls this man 'a *kaleidoscope* equipped with consciousness'. Whereas Poe's passers-by cast glances in all directions which still appear to be aimless, today's pedestrians are obliged to do so in order to keep abreast of traffic signals. *Thus technology has subjected the human sensorium to a complex kind of training.*

The question here is how in the post-World War I period technology subjected the sense of hearing and the practice of listening to this complex kind of training. It is not just that people were hearing new sounds—often sounds made by machines, and sometimes machines designed to produce sound. As the second quotation from Musil suggests, people were also beginning to think about hearing in new ways.[5]

Although I emphasise the new ways of hearing described and enacted by writers and musicians, the early decades of the twentieth century were a time when scientists and technologists were also rethinking the mechanics and meaning of sound along similar lines. Acousticians in the 1920s would have thought about the link between sound and a specific place less in terms of the audible character or aural experience of a place—Musil's exile returning to hear an unmistakable Vienna, for instance—than in terms of the reverberation that constitutes the acoustic signal of a particular space or place. And they would have thought of it as a problem insofar as it was seen as an impediment to the clarity of sound as signal or information. The aim was to make sound clear, direct and functional. This meant making it less reverberant, thus divorcing sound from place and making spaces sound increasingly alike. This principle applied equally to the emerging technologies for the reproduction of sound: the telephone, the gramophone and radio. Here again the aim was efficiency: faithfulness in reproducing an inevitably fuzzy and unfocused original was less important that stripping the reproduced sound of extraneous noise and interference so that a hearer or listener could easily decode and understand it.[6]

The machine age

In his documentary novel, *The Life of the Automobile* (1929), Ilya Ehrenburg describes the sounds of the Citroën factory outside Paris in a way that harks back to Simmel's image of the individual as a cog, forward to Charlie Chaplin's satire on the assembly line in *Modern Times* and tangentially to the concerns of acousticians.

> The Citroën works had twenty-five thousand employees. Once, they had spoken different languages. Now they kept silent. A close look revealed that these people came from different places. There were Parisians and Arabs, Russians and Bretons, Provençals and Chinese, Spaniards and Poles, Africans and Annamites. The Pole had once tilled the soil, the Italian had grazed sheep, and the Don Cossack had faithfully served the Tsar. Now they were all at the same conveyor belt. They never spoke to one another. They were gradually forgetting human words, words as warm and rough as sheepskin or clods of freshly plowed earth.
>
> They listened to the voices of the machines. Each had its own racket. The giant drop-hammers boomed. The milling machines screamed. The boring-machines squealed. The presses banged. The grinding-lathes groaned. The pulleys sighed. And the iron chain hissed venomously.
>
> The roar of the machines deafened the Provençals and the Chinese. Their eyes became glassy and vacant. They forgot everything in the world: the color of the sky and the name of their native village. They kept on tightening nuts. The automobile had to be noiseless. Engineers sat and thought. How could they build a mute engine? These valves had to be silenced. The buyer was so nervous! The men along the belt had no nerves. They only had hands: to tighten a nut, to fasten a wheel.[7]

This passage brings together a number of features in the modern soundscape. First, and most obvious, there is the brutal, deafening sound of the machinery of mass production. This is what people working in factories would have heard. Second, however, the passage connotes ways of thinking about the fact that these workers were being subjected to this deafening and deadening environment. There is Ehrenburg's Marxist critique, in which the only concern for noise reduction is aimed at assuaging the sensitive ears of bourgeois car-buyers. In fact, as Ehrenburg implies, noise at this time was being seen increasingly as a problem in need of technical solutions—and for reasons rooted in the production process as well as in consumer culture. Excessive noise was recognised as a health hazard needing treatment by physicians, and one that could indeed have a negative impact on the productivity of workers. Engineers and technologists were therefore developing new electro-acoustical instruments to measure and quantify sound with a view to mitigating its undesirable side-effects. Thirdly, there were increasing efforts to act on the material and social causes of noise. As noise pollution became an issue, noise abatement emerged as a significant progressive and/or conservationist political cause, engineers sought ways to make machines quieter and architects paid unprecedented attention to sound proofing and the acoustic design of buildings. Such problems, and the solutions being sought to them, reinforced the sense of noise as something independent from time and space, as decontextualised signals hitting the ear rather than the sort of unique audible dimension of a place that Musil heard in Vienna.

Ehrenburg's description of Citroën's workforce adds a fourth element, which can be found in many contemporary accounts of the soundscape of the 1920s. This is the image of the modern metropolis as the new Babel, as the number of languages to be heard in the European capitals multiplied. Here, Ehrenburg lists this linguistic diversity only to note how it is silenced, drowned out by the desensitising machines. In doing so, he is pointing to a paradox of capitalist development. The type of mass production represented by Citroën acted as a magnet for people, attracting them from around the world to the great industrial centres of Europe and America. As it did so, however, it obliterated their specific differences—as it also tended to do through its colonising appropriation of large parts of the globe.

The themes of migration, standardisation and atomisation again recall Simmel's account of the impact of the modern metropolis on subjective life. The overload of external information and stimuli in the modern metropolis, he believed, had provoked a new armouring of the modern self evident in a blasé attitude towards others and a purely aesthetic individualism. That adaptation of metropolitan stimuli to the ends of self-creation can be seen in the embrace of Berlin's noise and mechanisation by the novelist Alfred Döblin, author of *Berlin Alexanderplatz*: 'The city as a whole has an intensely inspiring, energizing power; this commotion of the streets, shops and vehicles provides the heat I must have in order to work, at all times. It is the fuel that makes my motor run.'

The defensive response, in contrast, is exemplified tellingly in a 1926 article, 'The Blue-Black Rose', by the philosopher and social critic Theodor Lessing. He found modern Berlin just too noisy: 'I hate the cries of the street merchants and newspaper vendors. I hate the ringing of the church bells, I hate the senseless noise of the factory sirens, but what I hate most are the stinking autos.'

For Lessing, as for Schopenhauer with his diatribes against the infernal cracking of whips in the nineteenth century, the sounds of the city and its machines represented something worse than an intrusion on his right to privacy. Their noise destroyed the silence necessary for inner reflection, thought and so self-development and artistic or intellectual creation. Driven to distraction, Lessing fantasised desperate measures.

> Banish the 'symbols of culture' from the thoroughly wired landscape, filled with advertisements and smokestacks; perhaps one site will remain holy and unspoiled. I've decided to steal a pocketwatch at some point—with the hope of being arrested. In prison I will at least have peace from the rug beating, the piano playing, the car horns, the gramophones, and the telephones.

In his irritation and distraction, Lessing does not discriminate between different kinds of noise: long familiar interruptions such as church bells, rug beating and piano playing, the side-effects of modernisation such as factory sirens and cars with their horns, and machines designed specifically to produce sound as a means of communication (such as the telephone) or a commodity (such as the gramophone). He therefore fails to notice, or to address, what Benjamin saw in technologies such as the telephone or gramophone. They did not simply disrupt old habits of perception and selfhood; through their complex training of the senses, they reconfigured ways of relating to oneself, to others and to the world.

As many historians and critics have observed, a key feature of this modern experience was a new dynamic of time, space and presence, the transcendence of distance. Looking back on the achievements of his fellow acousticians in 1929, for example, Harold Arnold, a physicist who had helped to develop techniques that improved the sound quality in telephones, radios and gramophones, declared: 'Now with one broad sweep the barriers of time and space are gone.'[8] The new machines created the possibility of virtual communication: talking to people not physically present, for example, or hearing music recorded long ago or far away. This world-shrinking and experience-expanding capacity of radio is invoked by Ehrenburg in an early section of *The Life of the Automobile* to frame his portrait of the first man to be killed in an automobile accident in France. The victim is presented as addicted fatally to the new experiences offered by new machines and technologies—not just radio, as here, but automobiles and cinema.

> Then a new serpent moved into his home. It hissed sweetly. It was the radio. Bernard's days still had an appearance of well-being. But at night he went crazy. He wore warm slippers with pompoms. But he wasn't sitting at his fireplace; no, he was whizzing through the world. His lips moved suspiciously. He was looking for waves. Here was Barcelona...Here was Karlsruhe...The German word 'bitte.' Bach. Spaniards. A charleston. The winner of the race at Oxford. The Royal Dutch rates. An Italian lesson: forte, morte, cannelloni. The victory of the Conservatives in Sweden. The bells of the Kremlin: The *Internationale*. Another charleston. The world moohed, bleated, meowed.

The 1920s were the decade of radio. The British Broadcasting Company (BBC) began its broadcasts from Savoy Hill on the evening of 17 September 1922, to be followed on 6 November by the first private radio transmitter in France—called initially Radiola but soon renamed Radio Paris—and then a year later, on 29 October 1923, by the Vox record company's broadcasts from Berlin. Ehrenburg seems to have been right to stress the cosmopolitanism of the content—a cosmopolitanism born largely of necessity, as radio stations were forced to broadcast whatever appropriate material they could lay their hands on. Here is the *Illustrated London News*'s description of the type of listening offered in the BBC's earliest days: an afternoon concert from the Eiffel Tower, a Tuesday evening concert from the Marconi works at Chelmsford, Thursday evening concerts from the Hague and the Sunday afternoon 'Dutch concert'.[9] The mastery of nature and the annihilation of time and space celebrated by Harold Arnold were not, however, just technical breakthroughs. The ability of film to bring images of the world to a cinema near you and the power of radio to make sounds from around the world available to you in your home—dressed, if you liked, in your slippers with pompoms—were major factors in the appeal of the two media. They seemed to promise a new universalism transcending differences of language and culture.

Radio was not just cosmopolitan in its content. It also created a new audience, a new community, that was at least potentially universal in the sense that, however widely scattered it may have been across cities and nations, it was nonetheless linked together in time 'with a simultaneity that made physical travel seem antediluvian by comparison'.[10] Paradoxically, however, this experience of hearing the world in your living room did not lead to more extensive or less constrained communication. Instead, it made people conscious that where they happened to live was not the world, but one place in an increasingly complicated and mysterious world. It emphasised the atomisation and aesthetics of defensive self-creation diagnosed a couple of decades earlier by Simmel. The virtuality of radio also added an uncanny dimension to the objectification of modern life; a sense of placelessness as everyday experience was mediated increasingly through the sounds and stories carried electronically through the air. In 1928 Berlin, Mischa Spoliansky and Marcellus Schiffer wrote a song, *Es liegt in der Luft* (*There's Something in the Air*), which captures something of the novelty and the paradox of this new media environment. First, they notice how telephone, radio and cinema 'informationalise' culture, turning it into something like electricity, into signals even more than signs.

> There's something in the air called objectivity (*Sachlichkeit*),
> There's something in the air like electricity…

But then they go on to conjure up the disconcerting, incredible barrage of sounds and images being carried by those signals.

What has come over the air these days?
Oh, the air has fallen for a brand new craze.
Through the air are swiftly blown
Pictures, radio, telephone.
Through the air the whole lot flies,
Till the air simply can't believe its eyes.
Planes and airships, think of that!
There's the air, just hear it humming!
Trunk calls, Trios in B flat
In the gaps that are left a picture's coming.

This was a strange and confused new world, an immaterial yet compelling soundscape populated by mechanically reproduced voices and sounds that changed the very nature of subjective being. 'As society becomes progressively aestheticized,' writes Michael North, 'as audiences begin to consume imaginative and symbolic materials as they had previously consumed material goods, then everyday life acquires *an inherently ironic distance from itself*.'[11]

Marcel Proust was an astute observer of the way in which new media and communication technologies produced this experience of internal distance. In *The Guermantes Way*, published in 1920 and 1921, for example, the narrator, Marcel, is thrown by his first telephone conversation with his beloved grandmother, as the 'tiny' and 'abstract' sound of her voice, divorced from her face and her physical presence, is translated into pure signal.

It is she, it is her voice that is speaking, that is there. But how far away it is!…

Many were the times, as I listened thus without seeing her who spoke to me from so far away, when it seemed to me that the voice was crying to me from the depths out of which one does not rise again…for always until then, every time that my grandmother had talked to me, I had been accustomed to follow what she said on the open score of her face, in which the eyes figured so largely; but her voice itself I was hearing that afternoon for the first time.

For Marcel, the uncanny experience of hearing this disembodied voice acts as an intimation of mortality, accentuating as it does the finitude of the body.

'Granny!' I cried to her, 'Granny!' and I longed to kiss her, but I had beside me only the voice, a phantom as impalpable as the one that would perhaps come back to visit me when my grandmother was dead. 'Speak to me!' But then, suddenly, I ceased to hear the voice, and was left even more alone…It seemed to me as though it was already a beloved ghost that I had allowed to lose herself in the ghostly world, and, standing alone before the instrument, I went on vainly repeating: 'Granny! Granny!' as Orpheus, left alone, repeats the name of his dead wife.[12]

The uncertain borderline between life and death also provokes Leopold Bloom to reflect on the uncanniness of another sound technology in James Joyce's *Ulysses* (1922). This time it is the gramophone, which Bloom speculates about as a means of recording and perpetuating the human voice.

> How many! All these here once walked round Dublin. Faithful departed. As you are now so once were we.

> Besides how could you remember everybody? Eyes, walk, voice. Well, the voice, yes: gramophone. Have a gramophone in every grave or keep it in the house. After dinner on a Sunday. Put on poor old greatgrandfather. Kraahraark! Hellohellohello amawfullyglad kraark awfullygladaseeagain hellohello amawf krpthsth. Remind you of the voice like the photograph reminds you of the face. Otherwise you couldn't remember the face after fifteen years, say.[13]

In *The Wasteland*, also published in 1922, T. S. Eliot invokes the more usual use of the gramophone as a music machine in the passage about a typist's casual sexual encounter with a 'small house agent's clerk…on whom assurance sits / As a silk hat on a Bradford millionaire'. Here it is not death that is summoned up but rather the deadening of affect in modern life. The recorded music in the typist's squalid flat conveys not just Eliot's disgust at her mechanical and disengaged coupling, but more generally his disdain for the dehumanised, automaton-like quality of contemporary culture.

> She turns and looks a moment in the glass,
> Hardly aware of her departed lover;
> Her brain allows one half-formed thought to pass:
> 'Well now that's done: and I'm glad it's over.'
> When lovely woman stoops to folly and
> Paces about her room again, alone,
> She smoothes her hair with automatic hand,
> And puts a record on the gramophone.

A thread running through these various accounts of mechanically reproduced sound is their sensitivity to the double reality of an emerging post-war world. On the one hand, the noise of modern life and the technologies of sound reproduction are part of a changing external reality: Simmel's crystallised spirit, Musil's durable stuff or simply 'objectivity'. The other reality, no longer conceivable in terms of individual human qualities, is subjectivity mediated through mass-produced sounds and technologies that dislocate or subvert the familiar coordinates of space and time. This learned estrangement of the senses transmutes into a disturbing rhythm that is sometimes stultifying and sometimes febrile. This is the inner reality and rhythm of modern life lived to the tempo of jazz.

The jazz age

In 1931, F. Scott Fitzgerald reflected on the era that he had named 'the Jazz Age'. The word jazz, he observed, 'is associated with a state of nervous stimulation, not unlike that of big cities behind the lines of a war'. This is a useful reminder that jazz operated in the 1920s not just as a loose name for a wide variety of musical styles, but above all as a term of art. People were certainly hearing new kinds of popular music, in music halls and nightclubs and especially as recorded on disc and broadcast on the radio. But jazz above all provided a metonym for talking about the times, as well as a soundtrack to them. In 1924, for example, Irving Berlin, who would have been considered a jazz composer, compared the music to the 'rhythmic beat of our everyday lives'. 'Its swiftness is interpretive of our verve and speed and ceaseless activity. When commuters no longer rush for trains, when taxicabs pause at corners, when businessmen take afternoon siestas, then, perhaps, jazz will pass.'[14]

Jazz was thus heard as the rhythm of modernity and as the noise of machinery. Jazz also mimicked the universal flow of capital and of commodities by spreading everywhere. A New York writer reported home from his world travels:

> No sooner had I shaken off the dust of some city and slipped almost out of earshot of its jazz bands than zump-zump-zump, toodle-oodle-doo, right into another I went. Never was there a cessation of this universal potpourri of jazz. Each time I would discover it at a different stage of metamorphosis and sometimes hard to recognize, but unmistakably it was an attempt at jazz.[15]

At the same time, jazz was part of the polyglot cosmopolitanism of Europe's capitals. Langston Hughes was working as a dishwasher in Le Grand Duc nightclub in Paris when he wrote his poem *Jazz Band in a Parisian Cabaret* in 1924.

> Play that thing, Jazz band! Play it for the lords and ladies, For the dukes and counts, For the whores and gigolos, For the American millionaires, And the school teachers Out for a spree. Play it, Jazz band! You know that tune That laughs and cries at the same time. You know it. May I? Mais oui. Mein Gott! Parece una rumba. Play, jazz band! You've got seven languages to speak in And then some, Even if you do come from Georgia. Can I come home wid yuh, sweetie? Sure.[16]

Although universal and cosmopolitan, jazz was also heard as specifically American and in some often vague way as black. Kurt Weill in 1926 was more musically precise than many of his contemporaries:

> The rhythm of our time is jazz. The Americanization of our whole external life, which is happening slowly but surely, finds its most peculiar outcome here. Unlike art music, dance music does not reflect the sense of towering personalities who stand above time, but rather reflects the instinct of the masses…Negro music, which constitutes the source of the jazz band, is full of complex rhythm, harmonic precision, and auditory and modulatory richness which our dance bands simply cannot achieve.[17]

Being heard as American and black, jazz embodied the very sound of the crisis as which modernity was experienced in Europe. As early as 1921, the Bloomsbury art critic Clive Bell hoped that the fad for jazz was passing, because it represented for him banal and immature tendencies in modernity rather than the critical edge of modernism that he saw in the paintings of Picasso, Braque and Derain. The Paris-based German commentator Ivan Goll also picked up on this anxiety, but with a greater degree of self-awareness. Like many people formed by the trauma of the war, he saw that the vogue for jazz and for primitivism generally signified a loss of confidence in European culture and the desperate desire for a new cultural authenticity. In his dispatches to *Die literarische Welt* from the cultural front, he alerted Berliners that Josephine Baker and the *Revue nègre* were on their way.

> The Negroes are conquering Paris. They are conquering Berlin. They have already filled the whole continent with their howls, their laughter. And we are not shocked, we are not amazed: on the contrary, the old world calls on its failing strength to applaud them…One can only envy them, for this is life, sun, primeval forests, the singing of birds and the roar of a leopard, earth.

Goll was, however, alert to the parodic nature of much of *La Revue nègre*, and scoffed at the assumption that its version of jazz and black America had anything to do with Africa.

> These Negroes come out of the darkest parts of New York. There they were disdained, outlawed; these beautiful women might have been rescued from a miserable ghetto. These magnificent limbs bathed in rinse water. They do not come from the primeval forests at all. We do not want to fool ourselves. But they are a new, unspoiled race.

As Goll continues, he articulates a characteristic 'cultural eugenics'—the idea that an exhausted and etiolated European culture needs reinvigoration by 'Negro' blood.

> Their dancing comes out of their blood, their life…The main thing is the negro blood. Drops of it are falling on Europe—a land, long dry, that has almost ceased to breathe. Is this the cloud that looks so black on the

horizon? A shimmering stream of fertility?…Do the negroes need us, or do we not rather need them?

Whereas Bell railed against jazz because it confounded hierarchies of aesthetic cultural value and threatened to make 'the coloured gentleman who leads the band at the Savoy' the ultimate arbiter of taste, Goll heard in the sound of jazz and saw in the movement of black bodies a grotesque image of the strength, authenticity and intensity of affect that had been damaged mortally by the Great War and that were being further undermined by the American culture of which commercial jazz was another product.

The musician Paul Stefan put this more programmatically in an editorial for the avant-garde music journal *Anbruch* in 1925.

> For us, jazz means: a rebellion of the people's dulled instincts against a music without rhythm. A reflection of the times: chaos, machines, noise, the highest peak of intensity. The triumph of irony, of frivolity, the wrath of those who want to preserve good times. The overcoming of Biedermeyer hypocrisy.[18]

The impact of jazz on Europe in the 1920s was, of course, more complex and more nuanced than I have been able to convey here. I have simply sketched how and why jazz functioned as the sound of modernity, and how it helped to school an emerging mentality. This had more to do with the way the music was heard in Europe than with the reasons why it was created in America, although the imagined relationship between Europe and America was a significant part of it. The actual cultural significance and aesthetic value of jazz were always greater than its detractors feared and often less than its supporters hoped. It was not yet a music of conscious African-American self-expression: it was the sound of an ambivalence towards the fact of modernity, a representation of Benjamin's complex training of the senses but also a sometimes startlingly inventive commentary on it.[19]

Sound of the city/dissonant modernism

In 1922, John Cournos, a Russian-born Jewish American shuttling between New York and London, published a novel that attempted to capture the rhythms and dislocations of the modern metropolis. *Babel* is in large part a *roman à clef* that (rather like *The Man Without Qualities*) looks back from a post-war perspective to London in 1913. It starts with a reflection on the apparently universal spread of capitalism and mechanisation and their links to imperialism. The narrator, Gombarov, soon begins to question whether this universalism is desirable or sustainable. When he first arrives in London from New York, he is first struck by the distinctiveness of the city's sounds as he boards a bus at Victoria Station. In contrast with the angular, wiry abstraction of Musil's Viennese soundscape, Cournos creates something like a post-impressionist tone poem.

Having left Westminster Cathedral behind, the red 'bus almost noiselessly glided on, cunningly brushing past other red 'buses, with a gondola-like grace which was incredible; at times no more than a hair's breadth separated them. And there was no sudden, sharp, shrieking noises of taxi-horns and overhead trains as in New York; but there was a trembling and a rumbling in the air, steady and constant, the even breathing of modern life over vast spaces. All the noises were swallowed up and became as one noise, vibrant like that of a ship's turbine, incessantly throbbing, reduced to normal pulsation, diffuse mellowness of a tone painting, in which conflicting colours take their place without quarrelling with one another, and none shrieking. This, Gombarov had time to observe before reaching the end of his journey, had its counterpart in the physical contours of the streets, which were curiously free from sharp abutting angles so characteristic of the streets of the New World.[20]

Sitting on the top deck of the bus, Gombarov begins to grasp the intimate dynamic between the standardisation of the external world and the atomisation of subjective life—again, a theme of Simmel's— as he notices the cosmopolitan entertainments on offer along the Charing Cross Road.

The soul of old England was left behind in Trafalgar Square; the 'bus rolled on through one of the corridors of the new England. Up Charing Cross Road, past a cinema house, announcing 'The Grim Avenger: A Thrilling Romance of Three Continents', past the Hippodrome, blazing with lights; past the buildings of new flats, utterly banal but for the curve of the old street; past a music hall, flaunting across its front the pirouetting figure of a Russian toe dancer on a coloured screen, while underneath, flashing for the world to see, letters of bright light proclaiming other attractions: a Cockney Comedian, a Spanish Tango Turn, a Swedish Acrobat Troupe, American Clog Dancers, an Argentine 'Stunt' Artist, Naughty Fifi the French Comic Chanteuse, and Mimi, her Eccentric Accompanist, and so on, and so on: 'How amazingly international!' mused Gombarov, and laughed to himself, as the after-thought struck him: 'And here am I, a Russo-American Jew, looking on!'

Was this chaos or unity? It was chaos, and had a unity after a fashion. It was the unity of a many-tuned medley, each tune of which maintained its entity, losing it only at the moment of embracing another tune; at best, it was the unity of ultra-modern music, shaped out of discords, beaten but not molten into a harmony.[21]

This play on musical chaos and unity to capture the social complexity of the modern metropolis and the subjective life of its citizens provides a suitable metaphor to draw together the collage of quotations through which I have tried to recapture what a European capital in the 1920s might have sounded like, and how that noise might have been heard. Without, I hope, conflating the question of sound with the question of music, I have tried to show how modern music, primarily jazz, not only formed part of the new soundscape but also tried to make sense (or at least art) of it.

Through this evocation, however, I have also been making an argument about the sensual, preconscious dimension to the way in which historical change is experienced. Mass industrial production, consumer culture, post-war reconstruction and adaptation, the rise of America, a faltering democratisation, media technologies—all these powerful forces, and more, were experienced and to some extent *made sense of* through the *senses*, including the sense of hearing. Meaning, it has been said, occurs when sound meets prejudice.[22] 'Prejudice' is a nicely chosen word. Less technical than Raymond Williams' 'structure of feeling' or Pierre Bourdieu's 'habitus', it connotes something less definite and certainly less easily describable even than the shared stories, myths and images that constitute what we have come to think of as 'cultures'. It suggests something more visceral and more elusive than that: unspoken and even unthought frames of perception, intuition, expectation and constraint that are nonetheless saturated by memories, desires and fears. These frames—subject to the complex training of technology and objective culture—are the way we hear and see the world, and thus the way we live history.

I have tried to indicate how strange was the modern sensory world coming into being in the 1920s. Even as scientists tried to remove extraneous distractions from recorded sound, the imaginative space of everyday life was becoming increasingly and uncannily over-populated by mechanical sounds and disembodied voices. The virtualisation of modern existence—accelerated if not caused by media technologies—could well have contributed to a characteristically modern sense of placelessness and even homelessness. This nostalgia is captured in David Vogel's description of the look and sound of Vienna at dusk in his 1929 novel, *Married Life*.

> In the mild spring air a pure, gentle stillness seemed to drop from the darkening sky. The deserted streets looked as if they had just been swept. The city was sinking into sleep in the orange glow of the streetlamps. From time to time, at increasing intervals, a tram split the silence like a nightmarish awakening. A distant train emitted a long, muffled hoot. And for a moment the imagination was captured by long journeys through the soundlessly breathing night, strange cities populated by millions of human beings.[23]

ENDNOTES

[1] I use 'soundscape' in the sense defined by Emily Thompson: 'Like a landscape, a soundscape is simultaneously a physical environment and a way of perceiving that environment; it is both a world and a culture constructed to make sense of that world. The physical aspect of a soundscape consists not only of the sounds themselves, the waves of acoustical energy permeating the atmosphere in which people live, but also the material objects that create, and sometimes destroy, those sounds. A soundscape's cultural aspects incorporate scientific and aesthetic ways of listening, a listener's relationship to their environment, and the social circumstances that dictate who gets to hear what.' See Thompson, Emily 1988, *The Soundscape of Modernity: Architectural Acoustics and the Culture of Listening in America, 1900–1933*, MIT, Cambridge, pp. 1–2.

[2] Musil, Robert 1995, *The Man Without Qualities*, trans. Sophie Wilkins, Alfred A. Knopf, New York, p. 3.

[3] Ibid., p. 4.

[4] 'The individual has become a mere cog in an enormous organisation of things and powers which tear from his hands all progress, spirituality, and value in order to transform them from their subjective form into the form of a purely objective life. It needs merely to be pointed out that the metropolis is the genuine arena of this culture which outgrows all personal life. Here in buildings and educational institutions, in the wonders and comforts of space-conquering technology, in the formations of community life, and in the visible institutions of the state, is offered such an overwhelming fullness of crystallised and impersonal spirit that the personality, so to speak, cannot maintain itself under its impact.' Frisby, David and Mike Featherstone (eds) 1998, *Simmel on Culture*, Sage, London, p. 184.

[5] There is a chicken-and-egg danger here, pointed out by Jonathan Sterne. Developing a genealogy that 're-places sound reproduction within the longer flow of sound history', he argues that it is less that technologies of sound reproduction as such created new ways of hearing than that social techniques of listening that emerged in the course of the nineteenth century determined the construction of sound reproduction as a practice. Sterne, Jonathan 2003, *The Audible Past: Cultural Origins of Sound Reproduction*, Duke University Press, Durham, pp. 95–106.

[6] Thompson, *Soundscape*, p. 3.

[7] Ehrenburg, Ilya 1976 [1929], *The Life of the Automobile*, Urizen Books, New York, pp. 22–3.

[8] Quoted in Thompson, *Soundscape*, p. 320.

[9] North, Michael 1999, *Reading 1922: A Return to the Scene of the Modern*, Oxford, New York, p. 16.

[10] Ibid., pp. 15–16.

[11] Ibid., p. 208; emphasis added.

[12] Proust, Marcel 1968, *Remembrance of Things Past*, vol. 2, trans. C. K. Scott Moncrieff and Terence Kilmartin, Chatto & Windus, London, pp. 135, 137. For an insightful discussion of the Proust and Joyce examples, see Danius, Sarah 2002, *The Senses of Modernism: Technology, Perception and Aesthetics*, Cornell University Press, Ithaca, on Proust: pp. 14–15.

[13] Quoted in Danius, *The Senses of Modernism*, p. 182.

[14] Quoted in Thompson, *Soundscape*, p. 144.

[15] Burnet Hershey's article in *New York Times Book Review and Magazine* is quoted in Johnson, Bruce 2003, 'The jazz diaspora', in Mervyn Cooke and David Horn (eds), *The Cambridge Companion to Jazz*, Cambridge University Press, Cambridge, p. 34.

[16] Quoted in Edwards, Brent Hayes 2003, *The Practice of Diaspora: Literature, Translation and the Rise of Black Internationalism*, Harvard University Press, Cambridge, Mass., p. 63.

[17] Weill, Kurt 1995, 'Dance Music', in Anton Kaes, Martin Jay and Edward Dimendberg (eds), *The Weimar Republic Sourcebook*, University of California Press, Berkeley, p. 597 (originally in *Der Deutsche Rundfunk*, 4, 14 March 1926, pp. 732–3).

[18] Quoted in Taylor-Jay, Claire 2004, *The Artist-Operas of Pfitzner, Krenek and Hindemith: Politics and the Ideology of the Artist*, Ashgate, Aldershot, p. 100.

[19] See also Johnson, 'The jazz diaspora', p. 50: 'Louis Armstrong might have *made* music of the twentieth century, but it was not he who *made it* the music of the twentieth century—for that we have to go to culturally and geographically diasporic spaces, Whiteman and white men, American Jews like Gershwin, Anglo-European commentators and the fans, readers, and musicians who were in some ways guided by them. Especially in the 1920s, jazz was to a significant degree the invention of diasporic discourses and practices, of non-African-Americans.'

[20] Cournos, John 1923, *Babel*, Heinemann, London, p. 70.

[21] Ibid., pp. 73–4.

[22] Johnson, James H. 1996, *Listening in Paris: A Cultural History*, University of California Press, Berkeley, p. 2. Johnson is paraphrasing Hans-Georg Gadamer.

[23] Vogel, David 1989, *Married Life*, trans. Dalya Bilu, Grove Press, New York. pp. 29–30.

3. Speech, Children and the Federation Movement

Alan Atkinson

It is still something of a mystery as to why voters in Australia agreed to the federation of the colonies in 1901. The prevailing argument among historians for many years made much of regional economic interests, but during the 1990s a new type of Australian nationalism led to a focus on motivation of a more altruistic kind. The inspirational language used by men such as Edmund Barton and Alfred Deakin, by patriotic writers of fiction and verse and by oratorically gifted members of the Australian Natives Association (ANA), during the 1880s and 1890s, has been dissected for what it says about genuine feelings of Australian nationhood. John Hirst has argued that Federation was due to a real and widespread sense of shared territory—a land 'girt by sea'—and of common destiny.[1] Bob Birrell has likewise written in convincing detail about a large program of innovative and high-minded social reform, which he says was sketched out before 1901 by leading writers and activists and largely carried through thereafter, in the years before World War I.

Birrell's contribution is especially interesting for present purposes because he describes Federation not only as a business of sentiment (which is Hirst's main idea), but as a scheme of ideas, a challenge to contemporary imagination. He describes the proselytising work of the ANA, for instance, as an exercise in cultural self-improvement for members and for those who fell under their influence. Feelings of national involvement were understood by the ANA to be an inevitable result of better knowledge—of a deeper and wider comprehension of Australian circumstances.[2] I am not aiming here to argue around the point as to why Australians voted as they did. Only during the 1890s was there much widespread popular enthusiasm for Federation—leading to the successful referenda of 1899–1900—and the material I am using relates mainly to the 1870s and 1880s. I do want to make use of the idea that Federation was an effort largely of intellect and imagination—if I can use those words without suggesting that effort was justified in every detail. Benedict Anderson talks about nationalism as an ideology that depends on the existence, for each citizen, of an 'imagined community' to which he or she seems to belong. The imagining of that community I take to be an intellectual process, perfectly abstract and even counter-intuitive. It might be imbibed by adult reading—of newspapers in particular—but also, of course, it depends on what happens in schools.

The older approach to the federation movement, focusing solely on economic self-interest, seems now to have been partly discredited, but in one respect, perhaps, it was more firmly based than the current orthodoxy. It has become common practice since the 1990s to talk about Federation just as Americans think about the revolution of 1775–83. It is understood to be an event bearing immediately on us as a matter of national and individual identity. It therefore seems pivotal, even providential, within long-term Australian experience. Previously, Federation was used by historians like any ordinarily complex event—mainly as a fund of evidence about the circumstances of the time. It was a means of commenting, for instance, on the early evolution of the labour movement or on the vicissitudes of free trade. It was treated much as we might treat, say, the rebellion in Sydney in 1808 against Governor Bligh, which had a continuing significance mainly as the source of well-documented argument about contemporary ideas and relationships. This is the approach I want to take to the federation movement, or rather to its early stages. Federation can be understood as the result of a particular cast of thought, a particular way of envisaging the world, which—instead of prefiguring current national and individual identity—was alien to the way we think now.

During the second half of the nineteenth century, in every one of the Australian colonies, the elementary school system—the system designed to cater for the mass of the population up to the age of 12 or 14—was reformed fundamentally. The changes imposed on teachers, parents and children were meant to meet two needs. In the first place, schools had to be provided for populations that were much larger and more scattered than they had been before the 1850s. Secondly, there were new ideas about the whole teaching process, which affected Australia as part of the British Empire and the European world. Schools were designed to create universal literacy and a mass culture for the nation-state. In addition, the 'art of teaching'—the way in which teachers were meant to behave in the classroom, and beyond—was largely altered.

The pedagogical reforms of that period have not been much examined by educational historians, at least compared with those of the early 1800s and the very end of the century. The federation movement has to be understood, however, in the context of orthodoxies prevailing in the 1870s and 1880s—methods and approaches that made their way into the schools, to great acclaim, mainly during the late 1860s and early 1870s. Under this regime, teachers made a self-conscious effort to improve on their predecessors. They tried to do more than impart knowledge: they also aimed to encourage independent reasoning. They were aware that children's minds developed in complicated ways and they understood that certain types of ideas matched certain stages of growth. Their purpose was 'the training of mental powers'. They tried to shape the thinking process, but also the training of mental powers was to be manifest in tangible ways. In theory, good teaching involved the child as a physical and intellectual being. Teachers, as an Australian school inspector said at the time, must 'train the eye to see, the ear to hear, and the hands to execute what the mind conceives and wills'.[3]

We are familiar today with pedagogical theory stressing individual and independent expression. This was not then a high priority. Expression was certainly important, but it was expected to show the obvious imprint of cultural authority. The basic subjects of the elementary curriculum were reading, writing and arithmetic. Reading was always listed first, and reading and writing were understood to be quite separate processes. In earlier generations, many children, especially girls, had never progressed beyond reading, and had never mastered either writing or arithmetic. Recent reforms were meant to give everyone a basic competence in all three and yet reading remained pre-eminent. 'No subject in our elementary school course,' said Thomas Burgan, a South Australian inspector of schools, in 1877, 'is perhaps more important than reading.'[4] All the evidence suggests that this was the common understanding. Besides the three Rs, a range of other subjects was available—at least in the bigger schools and to older children. Of these, as I say later, the most widely and effectively taught was geography, at least until about 1890 when, for reasons I will not explore here, it rapidly fell away. The mass teaching of geography was deeply characteristic of the period. It was also basic, as I will suggest, to the 'imagined community' of the nation.

Most of the detail below comes from the school inspectors' reports for the three leading Australian colonies from 1870 to 1890: New South Wales, Victoria and South Australia. I say 'leading' because in all three there was a cultural self-sufficiency, an independent sense of direction not to be found in Tasmania, Queensland or Western Australia. Inspectors had to give an annual account of the schools within their districts, comparing results from year to year, commenting on the physical condition of schools, equipment and furniture, the competence of teachers and the general behaviour of children. They commented also, if they saw fit, on the curriculum and on teaching methods used from place to place.

The length of the reports varied, but each might extend to several closely printed pages and all were published as part of the minister's annual report and bound with respective parliamentary papers. Part of each report was normally taken up with statistics, but generally the style was vivid and direct, and sometimes ironical, even whimsical, as inspectors struggled to reconcile the high aims of the school system with the realities they confronted from day to day. The contrast between official policy and humble experience runs parallel with the contrast I am trying to draw between imagined nationhood and what a Victorian inspector called 'the violent, boisterous talking-machine' to be found in many colonial schoolhouses.[5] The two realities can be seen face to face on a country road in Victoria's Gippsland in 1884, when Thomas Hepburn, the district inspector, met a weeping four-year-old as they both made their way towards a school up that day for judgement. 'On inquiring the cause of distress,' said Hepburn, 'I received the following answer, choked with sobs, "Oh! I don't want to go to school to-day, the *insect* is coming."'[6]

During the third quarter of the nineteenth century, the incidence of literacy (that is, the ability to read and write) increased to something like 90 per cent within each colonial population. At first sight, then, this was very much a literate culture, but there was something narrowly instrumental, as we would see it, about contemporary attitudes to literacy. It is clear from the evidence of the inspectors' reports that reading and writing were valued mainly as aspects of speech. They were contingent, in other words, on another medium. It is hard to find in the reports much reference at all to reading purely for its own sake. It was understood that books were rarely part of common life among the mass of the population; not many were to be found in the homes of the poor. Many idealists hoped that books would soon be everywhere and certainly the number available had increased vastly during the previous two or three generations. There was still no sense, however, that children might, through their reading, make for themselves a discreet world of knowledge and imagination, a kind of visionary alternative to the palpable world around them. The abstract information and ideas offered in books were valued as mental training. They prepared the way for respectability in this world and, perhaps, salvation in the next. They made men and women useful—to themselves and to others. They were a means of civilisation, but they were always merely a means. Books, and the material they contained, were rarely an end in themselves.

Some idea of the way in which reading was contingent on sound can be seen in the suggestion of one inspector that children should be persuaded that 'in reading they are actually listening to wise and intelligent men and women they have never seen'.[7] They were to do this mainly by reading out loud or, in other words, by listening to themselves. Reading out loud was the usual method of reading—or so it appears from the inspectors' reports. It was valued most highly at the elementary level—the level appropriate to the majority of the population—as a means of giving power and clarity to the voices of the children themselves. This was a very old idea. A seventeenth-century English pedagogue explained that children must learn about their own language, 'that they may be able in a gentle manner to converse with all sort of persons' ('gentle' can be taken to mean 'polite'). This early modern writer encouraged especially a concentration on words with 'a fair and ear-pleasing sound'.[8] The intimate link between reading and speaking was obvious still in late-nineteenth-century Australia. When John Kevin, inspector at Bega in New South Wales, wrote that 'the children in our schools…[are] not taught to read at all', he meant, with some exaggeration, that they were not taught to read well out loud. They knew their letters, but that was not enough. It was all a question of 'intonation and inflection of voice', and that, he said, must be 'natural, intelligent, and pleasing'.[9]

While the idea was old, however, the method was self-consciously innovative. Writing in 1871, James McCredie, inspector at Bathurst in New South Wales, remarked that in most schools in his district reading was still taught in outmoded ways:

> the names of the letters being mechanically dunned into the young child's mind, while nothing is said about the powers of the characters, and that they are mere marks to represent certain sounds. The pupil's interest is neither excited, nor is his intelligence cultivated by the irrational absurd method in question, and he is left to grope his way by dint of memory into a knowledge of the pronunciation of words, which he is made laboriously to spell over letter by letter before he names them.

This 'foolish custom', McCredie said, did more harm than good because it was hard afterwards for the child to understand that 'spelling and naming' were two very different things. It was naming—recognising words as a whole and saying them out loud—that was the first step of any importance in reading.[10] J. S. Jones of Armidale, New South Wales, made the same complaint at the same early stage: 'Few teachers in my district employ a judicious combination of the phonic and the look-and-say principles of teaching reading. Most of them adopt the alphabetic or spelling method, which is not only contrary to common sense, but is also uninteresting, tiring, and uncertain, to both pupil and teacher.'[11]

Parents were part of the problem, at least in South Australia, where they seem to have had unusual influence. They 'naturally cling', it was said, 'to the method of their early days, by which years were sometimes devoted to what was known as the A, B, C'.[12]

Educational historians have made too little of the complicated importance of speech for educational methods in this period. It jars a little with the common belief that late-nineteenth-century children were prepared partly for a world of silent reading and partly to be the passive employees of industry. In fact, the importance of speech training was stressed as a means of building up a truly democratic people. Charles Henry Pearson, in his report on the Victorian school system in 1878, spoke of the need to follow the example of the United States. There, he said, in the schools, '[m]ere children are taught to speak out so that they may be heard by a large number in a large room'. As a result, among adults in public meetings, 'the general fluency of comparatively untried speakers' was far beyond the normal standard in Britain and its colonies.[13] A Victorian inspector echoed this idea in 1882. 'In a community like ours,' he said, 'where one may be called upon at any time to speak before his fellow-men, it is an immense advantage for a young man to have early become accustomed to the sound of his own voice.' Once again, the United States—'a nation of ready public speakers'—ought to be the model.[14]

The matter went deeper than this: the kind of speech encouraged in schools was not only individual, it was communal. A good deal of class time was taken up with what was called 'simultaneous recitation': the chanting of lessons. It is easy today to think of this method of teaching as proof of a failure of effort and imagination among teachers. It was indeed justified partly as a means of keeping children occupied and their minds alert, but it was also part of the reforming program. 'A marked defect in the schools,' said J. S. Jones in 1871, 'is the almost entire absence of recitations.' By speaking and listening together, by taking advantage of what a leading British reformer had called 'the sympathy of numbers', children learnt to adapt their voices to the ideal. Simultaneous recitation was a means of induction into a uniform system of speech. Educational historians have featured the individual and competitive spirit, which was certainly part of educational practice in this period.[15] On the other hand, however, schools—like nations—were meant to be communities. They were meant to be established on a foundation of sameness. That sameness was manifest partly in voices—a point I will come back to at the end of this chapter.

I now turn to geography—the teaching of which, surprisingly perhaps, helps to explain the importance of voices in the schools. Geography was a discipline of enormous interest to this generation. New methods of printing—first implemented in the 1830s—meant that maps were now much more easily available than they had been before.[16] They began to appear in textbooks, although for most of our period it was usual for teachers to rely instead on large sheet maps hung up or draped over furniture. Children did not usually have maps of their own. In spite of being fairly common, then, maps were still at least slightly mysterious.

Cartographic literacy—associating space on a map with space in the real world—was and is a skill in its own right. It was a peculiarly fascinating skill at this time. It had obvious practical uses but also, like literacy proper, it seemed to be the key to other worlds. Unlike literacy proper, it led inevitably to silent rather than spoken imaginings. Flora Thompson, who had been an English schoolgirl in this period, recalled the maps hung about her classroom. 'During long waits for her turn to read [aloud],' she said (writing in the third person), '[she]…would gaze on these maps until the shapes of the countries with their islands and inlets became photographed on her brain. Baffin Bay and the land around the poles were especially fascinating to her.' For some children, on the other hand, such shapes were deadening. Ethel Richardson (afterwards Henry Handel Richardson) remembered her own school days in Melbourne. 'She knew,' she said (also using the third person), 'from hearsay, just how England looked…its ever-green grass, thick hedges, and spreading trees; its never-dry rivers; its hoary old cathedrals; its fogs, and sea-mists, and over populous cities.' This hearsay had nothing to do with the unfamiliar diagram she confronted in geography lessons: a map of the mother country 'seared and scored with boundary-lines, black and bristling with names'.[17] The trick lay in letting such shapes take on a many-layered but speechless imagery, as Flora Thomson did.

The profound way in which cartographic knowledge made its mark on contemporary imagination can be seen in an extraordinary book published in England in 1884, called *Flatland: A Romance in Many Dimensions*. The author, Edwin Abbott (writing as 'A Square'), was an Anglican theologian and schoolmaster. His story suggests a delight in geography and geometry. Thus, imagination struggles to comprehend worlds of different dimensions, from the one-dimensional to the four-dimensional and beyond. *Flatland* is remembered today partly as a precursor of computer graphics, or hypergraphics.[18] Its message, however, bears directly on the rise of mass cartographic literacy. Maps, after all, like computer graphics, render the familiar three-dimensional world in counter-intuitive two dimensions.

Rolf Boldrewood gives a neat account of the uses of geography for the individual in a story published in 1888–89, called *A Sydney-Side Saxon*. The hero starts by telling of his origins in poverty in England. As a boy, he says, he had read Walter Scott's story *Ivanhoe*. His imagination had fixed on Scott's description of a Saxon serf wearing a brass collar inscribed with details of his bondage to one lord and one place. For the hero, freedom from a metaphorical collar—from poverty and powerlessness—depended on knowledge, especially knowledge of geography. People of his rank in England, he says, 'knew no more about Australia, or Canada, or New Zealand, than the man in the moon', yet '[a]ny man or woman that can read and write, keep simple accounts, and understand a map, has got hold of the levers that move the world'. '[I]t is his own fault,' he says, 'if he doesn't prise out a corner for himself somewhere.' 'I was sharp about geography,' he goes on, 'so I looked out Australia, and found that there were divisions or colonies with large cities and houses, just like other places.' He came to this country, where he was taken up by a helpful squatter and shown a map of surrounding runs. There, concealed among lines that only the initiated could properly understand, was a sliver of uncharted territory, about seven miles by five, waiting to be taken up. On it, the hero made his fortune.[19]

For the 'imagined community' of the nation, and even more, perhaps, for the imagined community of race, cartographic literacy was crucial. In late-nineteenth-century debates about race, the powerful charm of geography is certainly very obvious, with its talk of expanding and retreating races and of 'black belts' and 'yellow belts' encircling the globe. The argument in Charles Pearson's celebrated book *National Life and Character*, published in 1893 shortly after he returned to England after many years in Australia, depended on this clear-cut sense—primitive as it seems today—of the spatial dimensions of humanity. So did contemporary nationalism.

Two distinct approaches were used in teaching geography in colonial schools. J. J. Fletcher has suggested that one method was inductive and the other deductive. The first, he says, 'began with the child's immediate environment and gradually moved out from there to include the world as a whole'. The second 'began with a view of the world (or even the universe) and worked down to particular areas'. The latter was, however, not wholly deductive because it rarely reached as far as the local and immediate.[20] In Victoria, where this top-down method prevailed until the late 1880s, the curriculum began in second class with '[t]he continents, oceans, and larger seas, with their relative positions', and the main point in third and fourth class was to add detail to this very large picture. In fifth class, there was a more particular focus, the work at that stage being described as '[m]aps of Europe and Australia'. Then in sixth class—the top class—the syllabus was as broad as could be: '[g]enerally of the world.'[21]

In South Australia, schools were required to use the inductive method. Teachers were told to concentrate to begin with on 'local geography'. This involved drawing spatial diagrams of the classroom, the school buildings and the neighbourhood and gathering knowledge of the surrounding landscape. When children had grasped the disciplinary method in this way they were shown maps on a larger scale and told of mountains, rivers and oceans, which they had never seen and could never hope to see, except in diagrammatic form. Inevitably in South Australia there was a strong emphasis on the colony and on the continent. In first class, children were taught '[c]ardinal points and local topography' and, in second class, they went on to '[t]he definitions of land and water and outlines of South Australia'. In third class, the curriculum was 'Australia in outline…South Australia in detail' and, in fourth class, '[o]f Europe…Ability to draw the maps of Australia and South Australia', while in fifth class, it was '[g]enerally of the world'.[22] As an inspector remarked, the child was first told 'something about his own neighbourhood, the hills that surround it, the creeks which flow through it, the seas which wash its shores…[a method] leading his mind forward from the known to the unknown'.[23] The main challenge lay in making the leap between dimensions, and from the concrete to the abstract. As one of the inspector's colleagues remarked, the advance 'from the geography of the neighbourhood to that of all South Australia is a great stride'. Maps were the key, because cartographic literacy applied in the same way to all places—large and small. This inspector believed, for reasons he did not explain, that the leap outwards from Australasia to the world was the hardest of all.[24]

Through most of the 1870s and 1880s, it was a prime goal for many inspectors—perhaps for them all—that children master the geography of the colony in which they lived. As James McCredie remarked in 1874, children ought to understand the physical characteristics of 'their own country'—and by that he meant New South Wales.[25] His colleague in the same colony, J. H. Murray, reported in 1883, 'I have endeavoured to impress upon teachers the idea that Australian boys and girls of twelve or thirteen years of age should be as thoroughly acquainted with the geography of their own country as European girls and boys are with theirs.' In spite of what this statement might imply to start with, what Murray meant was that a child in New South Wales should learn about New South Wales just as, say, a French child learnt about France.[26] In short, only 16 years before the referenda on Federation, geographical understanding was very different—among all ranks—from that on which Federation was to depend.

In geography, only towards the end of the 1880s do we find any interest in Australia for its own sake. In Victoria, the syllabus was changed at that point. As a result, according to an inspector, 'Victorian children are supposed to learn and to thoroughly know the leading features and the industrial resources of our own continent before they are taught much about the outside world'.[27] Therefore, he suggested, '[m]aps of the colonies should be supplied to all schools in lieu of maps of Asia, Africa, and America'.[28] This notion of 'our own continent' set against 'the outside world' was, of course, the essence of continental nationalism (although it was not clear yet that 'our own continent' included Tasmania). In the following year, C. H. Pearson, now Victorian Minister of Public Instruction, called for books to meet what he described as 'the particular needs of Australian children'.[29] The minister in New South Wales, J. H. Carruthers, agreed: 'To a large extent,' he said, 'the subject matter of the lessons should be Australian', or else it 'should be dealt with, as far as practicable, from points of view interesting to Australians'.[30] Such points of view must have been informed by the circumstances of daily life, wherever they seemed peculiar to the continent as a whole.

Inspectors were already beginning, very slowly, to make similar suggestions. In getting the children to read aloud and to memorise verse, said F. H. Rennick at St Arnaud in Victoria, in 1888, we should use 'our own Australian poetry'. 'Among the poems of Gordon and Kendall,' he said (and he later added Charles Harpur), 'we have specimens that breathe a true Australian spirit.' In our textbooks, for instance, information on 'Australian fauna and flora' ought to be mixed up with what he called 'gems of Australian song'.[31]

This leads me back to the question of reading, or rather, of speech. When inspectors commented on expression in reading they also commented, very often, on pronunciation. Problems with pronunciation had always been understood as a narrowly colonial rather than a continent-wide phenomenon. Many inspectors were interested in the way in which English regional peculiarities made their impact here. In Melbourne, A. C. Curlewis remarked on the way in which Victorians of various ethnic origins dropped their 'h's:

> From the Cockney and the Cornish the disease has spread to the Scotch and Irish. We [Victorians] are in a fair way of becoming a nation without an h in our vocabulary. From the average choir you will be pretty sure to hear ''eaven' and ''ell', and ''oly, 'oly, 'oly'. The average porter shouts out ''Awthorn' and ''Awksburn', and the average cabman 'Emerald 'ill'. There are other weak points in Victorian pronunciation; but this, I think, is the worst.[32]

His colleague, James Holland, wrote in the same terms. '[T]he distinguishing characteristics of Victorian pronunciation,' he said, 'are a slight tendency to a nasal accent and a want of crispness and distinctness in enunciation.' The Aborigines, he suggested, did the same.[33] It did not occur to him, apparently, in spite of his comparison with the Aborigines, that this might be a continental phenomenon. As with geography, knowledge of Victoria remained an end in itself. Issues of speech were anchored in Victorian society and Victorian space.

As with geography, however, ambiguities began to appear by the end of the period. Take the statement of Alfred Jackson, of Stawell, in 1890. 'Unless we have some decided improvement,' he said, 'the next generation must not be surprised if foreigners or neighbouring colonists are able to detect a Victorian rustic by his monotonous drawl.' On the one hand, Jackson divided non-Victorians into two categories: 'foreigners' and 'neighbouring colonists'. On the other hand, he still referred to common habits of speech in Victoria as if they were peculiar to that colony.

I have said that in South Australia the teaching of geography began with the local and moved by stages towards the global. The middle stage was the continent as a whole. This approach gave a kind of absolute value to the colony and the continent. In many ways, South Australians were always more willing than neighbouring colonists to see themselves as part of a community of provinces. They were more ready, for instance, than either Victorians or New South Welshmen to make formal and detailed comparisons between their school systems and those of their neighbours.

Similarly, in 1886, Alexander Clark, in South Australia, reported on his own district in such a way as to bring the continent automatically within his purview. 'Australians on the average,' he said, 'speak better English than the people of England themselves.'[34] What did he mean by 'better'? Naturally, the inspectors thought of their own usage as the best standard, but they placed particular stress on the need for uniformity from place to place, as if the two criteria amounted to the same. Speech was to be undifferentiated in every corner of the continent (perhaps in every corner of the Empire). It was to be contingent on the truths of geography and explored and controlled with the help of maps. A three-dimensional phenomenon—a phenomenon of the most impulsive, intimate and palpable kind—was to exist as well in two dimensions, on paper. The inspectors seem, in short, to have been confused as to where varieties of accent came from. On the one hand, they condemned what they called 'vulgarisms'—as if accent were a matter of social rank. On the other hand, and much more often, they spoke of 'colonialisms' and 'provincialisms'. They linked pronunciation—good and bad—with place.

Here, said Clark, 'the school children of Brisbane, Sydney, Melbourne, and Adelaide are much more nearly allied in speech than those of London, Plymouth, Leeds, and Newcastle'. He seems to have used an assumption that nation-building—whatever the nation might be—involved the breaking down of regional and local boundaries, a process that necessarily involved the wearing away of regional and local accents, plus strenuous efforts to make sure that no new peculiarities emerged as new communities became established. He had no hope, apparently, of a kind of providential unfolding of the Australian, or even South Australian, national type. On the contrary, as settlement became established, he thought, provincial differences would open up here as well. Teachers must work hard, therefore, to establish a single standard.[35] 'The Australias,' he said in 1888 (and by that he meant the various colonies of Australia), 'are at present devoid of the marked provincialisms of Yorkshire, Lancashire, or Somerset, and even of the less objectionable peculiarities of London or American speech.' 'Peculiarities' were, of course, the problem.[36]

Bad speech was a symptom of ignorance, of restricted horizons and of local attachment, when nation and empire were the true means of intelligent humanity. Bad speech was like the serf's brass collar in *Ivanhoe*—a token of bondage to be carried everywhere, binding its wearer to rank and place.

Here was a clear preference for continental nationalism, a belief in a continental type. Australia's imagined community was to be deduced from writing and from cartographic imagery and it was to be manifest in speech. Uniformity of speech was as telling, in other words, as uniformity of race. It might have been hard to agree on what that speech should be like. Leading writers of fiction—such as Henry Lawson, Joseph Furphy—might look to the language of the common man, to the vulgarisms condemned by inspectors. Their interest in the subject, no less vivid than the inspectors', shows that it was a matter of significant concern among a great variety of opinion-makers. Perhaps an interest in speech, as a symptom of national type, was unusually strong in this part of the world. Melvyn Bragg, in his account of the English language world-wide, remarks on Australians' distinctive 'love affair' with their own accent, beginning in these years. 'This was a people,' he says, 'finding its identity in the most essential and enjoyable way.'[37] It was a tortured love affair, no doubt, but some Australians were clearly happy to be bound to a place both limited and vast.

All these late-nineteenth-century authorities might well have agreed on the purpose of reading: on the need to make each child an effective citizen, civilised and efficient in the use of his or her voice. The great aim was also to create a uniformity of sound—parallel with uniformity of race—within the circumference of the nation.

ENDNOTES

[1] Hirst, John 2000, *The Sentimental Nation: The Making of the Australian Commonwealth*, Melbourne.

[2] Birrell, Bob 2001, *Federation: The Secret Story*, Sydney (earlier version published in 1995 as *A Nation of Our Own*), pp. 112–14.

[3] Burgan, Thomas 1882, 'Inspector's report, 1 February 1882, Report of the Minister controlling Education, 1881', *Proceedings of the Parliament of South Australia*, vol. 3, no. 44, p. 15.

[4] Burgan, Thomas 1877, 'Inspector's report, 23 January 1877, Report of the Minister controlling Education, 1876', *Proceedings of the Parliament of South Australia*, vol. 2, no. 34, p. 21.

[5] Roche, D. M. 1889, 'Inspector's report, 6 June 1889, Report of the Minister of Public Instruction, 1888–89', *Victorian Legislative Assembly Votes and Proceedings*, vol. 4, no. 98, p. 167.

[6] Hepburn, Thomas R. 1885, '9 April 1885, Report of the Minister of Public Instruction, 1884–85', *Victorian Legislative Assembly Votes and Proceedings*, vol. 4, no. 74, p. 151 (original emphasis).

[7] Whitham, C. L. 1880, 'Inspector's report, February 1880, Report of the Minister controlling Education, 1879', *Proceedings of the Parliament of South Australia*, vol. 3, no. 44, p. 15.

[8] George Snell (1649), quoted in Watson, Foster 1909, *The Beginnings of the Teaching of Modern Subjects in England*, London, pp. 40–1.

[9] Kevin, John 1883–84, 'Inspector's report, 11 January 1883, Report of the Minister of Public Instruction, 1882', *NSW Legislative Assembly Votes and Proceedings*, vol. 7, p. 718.

[10] McCredie, James 1870–71, 'Inspector's report, 23 February 1871, Report of the Council of Education, 1870', *NSW Legislative Assembly Votes and Proceedings*, vol. 4, p. 232.

[11] Jones, J. S. 1870–71, 'Inspector's report, 13 February 1871, Report of the Council of Education, 1870', *NSW Legislative Assembly Votes and Proceedings*, vol. 4, p. 213.

[12] Johnson, Edwin and James W. Allpass 1872, 'Inspectors' report, 25 January 1872, Report of the Council of Education, 1871', *NSW Legislative Assembly Votes and Proceedings*, vol. 2, p. 623; Hosking, James 1879, 'Inspector's report, n.d., Report of the Minister controlling Education, 1878', *Proceedings of the Parliament of South Australia*, vol. 3, no. 35, p. 22.

[13] Pearson, Charles H. 1878, *Report on the State of Public Education in Victoria and Suggestions as to the Best Means of Improving It*, Melbourne, p. 61.

[14] Lewis, Alexander T. 1882–83, 'Inspector's report, n.d., Report of the Minister of Public Instruction, 1881–82', *Victorian Legislative Assembly Votes and Proceedings*, vol. 3, no. 67, p. 215.

[15] Phillips, Derek 1988, The State and the provision of education in Tasmania, 1839 to 1913, Ph.D. thesis, University of Tasmania, pp. 321–6; Jones, J. S. 1870–71, 'Inspector's report, 13 February 1871, Report of the Council of Education, 1870', *NSW Legislative Assembly Votes and Proceedings*, vol. 4, p. 213; Johnson, William R. 1994, '"Chanting Choristers": Simultaneous recitation in Baltimore's nineteenth-century primary schools', *History of Education Quarterly*, vol. 34, pp. 14–16, 22.

[16] Coote, Anne 2004, Space, time and sovereignty: Literate culture and colonial nationhood in New South Wales up to 1860, Ph.D. thesis, University of New England, pp. 132–51.

[17] Thompson, Flora 1945, *Lark Rise to Candleford*, Oxford University Press, London, p. 174 (I owe this reference to David Kent); Richardson, Henry Handel 2001, *The Getting of Wisdom*, Clive Probyn and Bruce Steels (eds), University of Queensland Press, St Lucia, p. 66.

[18] 'A Square' [Edwin Abbott] 1884, *Flatland: A Romance of Many Dimensions*, Seely & Co., London; Banchoff, Thomas H. 2005, *From Flatland to Hypergraphics: Interacting with Higher Dimensions*, available from www.geom.uiuc.edu/~banchoff/ISR/ISR.html (accessed 19 April 2005).

[19] Boldrewood, Rolf 1925, *A Sydney-Side Saxon*, Sydney, pp. 12, 18, 32, 140–1.

[20] Fletcher, J. J. 1973, 'Teaching Geography in the National Schools of New South Wales 1848–1866', in S. Murray-Smith (ed.), *Melbourne Studies in Education*, Melbourne, p. 194.

[21] 'School curriculum, Ninth Report of the Board of Education, 1870', *Victorian Legislative Assembly Votes and Proceedings*, 1871, vol. 3, no. 33, p. 22.

[22] 'Standard of proficiency for scholars, Report of the Central Board of Education, 1873', *Proceedings of the Parliament of South Australia*, 1874, vol. 2, no. 24, p. 24.

[23] Jung, Emil 1875, 'Inspector's report, n.d., Report of the Central Board of Education, 1874', *Proceedings of the Parliament of South Australia*, vol. 2, p. 13.

[24] Dewhirst, Edward 1877, 'Report of the Council of Education, 1876', *Proceedings of the Parliament of South Australia*, vol. 2, p. 11; Dewhirst, Edward 1880, '19 February 1881, Report of the Minister controlling Education, 1880', *Proceedings of the Parliament of South Australia*, vol. 3, no. 44, p. 4.

[25] McCredie, J. 1873–74, 'Report of the Council of Education, 1873', *NSW Legislative Assembly Votes and Proceedings*, vol. 5, p. 328.

[26] Murray, J. H. 1883–84, 'Inspector's report, 31 January 1883, Report of the Minister of Public Instruction, 1882', *NSW Legislative Assembly Votes and Proceedings*, vol. 7, p. 711.

[27] Eddy, F. C. 1889, 'Inspector's report, 31 May 1889, Report of the Minister of Public Instruction, 1888–89', *Victorian Legislative Assembly Votes and Proceedings*, vol. 4, no. 98, p. 169.

[28] Eddy, F. C. 1888, 'Inspector's report, May 1888, Report of the Minister of Public Instruction, 1887–88', *Victorian Legislative Assembly Votes and Proceedings*, vol. 3, no. 99, p. 168.

[29] Pearson, Charles H. 1890, 'Report of the Minister of Public Instruction, 1889–90, 2 October 1890', *Victorian Legislative Assembly Votes and Proceedings*, vol. 3, no. 90, p. xxi.

[30] Carruthers, J. H. 1891–92, 'Report of the Minister of Public Instruction, 9 April 1891', *NSW Legislative Assembly Votes and Proceedings*, vol. 3, pp. 42–3.

[31] Rennick, F. H. 1888, 'Inspector's report, May 1888, Report of the Minister of Public Instruction, 1887–88', *Victorian Legislative Assembly Votes and Proceedings*, vol. 3, no. 99, p. 162; Rennick, F. H. 1890, 'Inspector's report, April 1890, Report of the Minister of Public Instruction, 1889–90', *Victorian Legislative Assembly Votes and Proceedings*, vol. 3, no. 90, p. 225.

[32] Curlewis, A. C. 1885, 'Inspector's report, 18 April 1885, Report of the Minister of Public Instruction, 1884–85', *Victorian Legislative Assembly Votes and Proceedings*, vol. 4, no. 74, p. 134.

[33] Holland, James 1887, 'Inspector's report, 30 May 1887, Report of the Minister of Public Instruction, 1886–87', *Victorian Legislative Assembly Votes and Proceedings*, vol. 3, no. 81, p. 142.

[34] Clark, Alexander 1886, 'Inspector's report, 16 March 1886, Report of the Minister controlling Education, 1885', *Proceedings of the Parliament of South Australia*, vol. 2, no. 44, p. 15.

[35] Ibid.

[36] Clark, Alexander 1888, 'Inspector's report, February 1888, Report of the Minister controlling Education, 1887', *Proceedings of the Parliament of South Australia*, vol. 3, no. 44, p. 27.

[37] Bragg, Melvyn 2003, *The Adventure of English: The Biography of a Language*, Hodder & Stoughton, London, pp. 281–86.

4. Sounds of History: Oratory and the fantasy of male power

Marilyn Lake

In my wildest egotistical dreams

In 1907, Arthur Atlee Hunt, the sensitive and self-regarding Secretary of External Affairs in the Prime Minister's Office, accompanied his 'Chief', Alfred Deakin, to the Colonial Conference in London. He reported on his experience, and that of the Prime Minister, in a series of confidential and detailed letters to his friend Bob Garran, in Melbourne, asking Garran to keep the letters as 'a memento of these important doings'.[1] Central to Hunt's reports to his friend were detailed accounts of Prime Minister Deakin's speech-making, widely praised as exemplary in its oratorical power.

What also becomes clear in Hunt's letters is the role of oratory as an indicator of masculine prowess in his Edwardian world. He was a keen analyst of the redeeming power of oratory and entertained fantasies on his own behalf. He wrote to Garran that he was sorely disappointed by Sir Wildred Laurier, Prime Minister of Canada, who was handicapped by 'a weak voice and a foreign accent'. He was surprised: 'Either he has faded from his former brilliancy, or else the standard of public speaking in the [Canadian] Dominion must be low, otherwise I cannot account for his reputation.'[2] He felt a nationalist pride in Deakin, who proved particularly impressive in the challenging role of after-dinner speaker. At the Pilgrims' Dinner, he was preceded by the Foreign Secretary, Sir Edward Grey, who 'in his curt, clear, incisive, cultured House of Commons manner was delightful to listen to, but there was no humour and no fire'.[3] In Deakin's speech, on the other hand, there were both: '[H]is sarcasm was polished and keen, and his enthusiasm was earnest and infectious.' Hunt judged it the best speech he had ever heard Deakin—or anyone else—ever make.[4] He was pleased to report to Garran that the Prime Minister's renowned ability as a speaker meant that his 'position as the most prominent and forcible of all the Colonial Premiers is greatly strengthened: not only is he infinitely the best speaker but he seems to be the only one who has the courage to say exactly what he thinks'.[5]

Deakin surpassed all other Dominion leaders in oratorical skill; the other speakers at the 1907 conference, Hunt noted, talked 'platitudes of the commonest type'. 'I have often felt I should like to have their chance,' he confided to Garran, 'and though in my wildest egotistical dreams I could never hope to approach even our Chief's poorest efforts…yet I am confident that within five minutes' notice I could have made a better speech than any other Colonial representative.'[6] In the cultured world of public men, oratory conferred power and prestige, hence their preoccupation and its place in their dreams.

Independent manhood

Deakin's oratorical triumph in London in 1907 was especially important in the political context. Despite the achievement of Federation, Australia remained constitutionally subordinate to British power. The contradiction between Australia's continuing condition of colonial dependency and its assumed status as a self-governing nation exacerbated continuing tensions between Deakin and the Colonial Office and made the conference a tense and testy affair. Deakin's biographer, J. A. La Nauze, observed that Deakin approached the Colonial Conference as one would a 'battlefield', the proud colonial ever ready to defend Australia's honour in the face of imperial condescension.[7]

Deakin yearned for recognition of Australia's independent manhood, a condition exemplified by the nation he liked to call 'the great republic'.[8] In literature as in life, he was attracted to American strength, vigour and power.[9] Harvard philosopher Josiah Royce enchanted him. After their momentous meeting in Australia in June 1888, Royce remembered Deakin in *Scribner's* as 'an admirer of America and of good scenery, a lover of life, of metaphysics, and of power'.[10] What we might call Deakin's republican desire, his love affair with American manhood, fuelled endless contests with the knighted Englishmen in the Colonial Office, who were loftily disposed, in their turn, to 'ignore Mr Deakin' and his litany of querulous complaints.[11]

A capacity for great oratory was one of the masculine virtues Deakin associated with the nation builders of 'the great republic'. He would have been pleased indeed to receive a letter from an American journalist comparing his speech-making at the Constitutional Conference in 1890 with the 'splendid oratory of Charles Sumner and Wendell Phillips'. Zadel Gustafson, who had heard him speak at the conference in Melbourne, in February that year, sent Deakin the article she had written, in which she compared his speech with the 'truth-charged burning words' of those great anti-slavery campaigners:

> It is with the memories of these men's lives and truth-charged burning words, fresh in mind that I write to say how great the delight I felt in hearing the Constitution and institutions of my country set before this Conference of Australasia in the very hour of the conception of the nation that is to be.[12]

Gustafson then analysed the elements that combined to render Deakin's speech into such powerful oratory. There were the cogent arguments that resulted from thoughtful study and investigation, his sympathy with his subject, the absolute fitness of the words chosen for his theme and the whole made luminous with 'the passion of Nationality'. As an American, Gustafson was especially moved by Deakin's admiration and knowledge of United States history, law and politics. 'If you can hold on your way with such lucidity, vigor, courage, discretion and passionate patience, as characterized your yesterday's speech, the union of these colonies will come,' she assured Deakin. And this historic event would occur largely through his efforts, which would also bring about closer relations with that 'great American Union of States' whose political character had been 'so clearly apprehended' by Deakin.[13] The young Victorian politician would have rejoiced in this tribute, though he was probably less pleased by Gustafson's description of him to her readers as 'tall, straight and dark like a rugged impression of Philip II of Spain'.[14] As an ardent Anglo-Saxonist, he would not liked to have been thought of as swarthy.

For Deakin, the Americans' oratory expressed their vigorous independent manhood and he would use his own oratorical power to redeem Australian manhood from its condition of colonial dependency, as he had done 20 years earlier in 1887, when he famously stood up to the British Prime Minister, the Marquis of Salisbury.

Prime Minister Deakin and Secretary Hunt had gone to the imperial capital in 1907, with the aim of challenging the 'limitations on our right of legislation' and 'removing from the Colonial Office practically all control of the self-governing Colonies'.[15] The British and Australian governments, he determined, must henceforth deal with each other as equals and as men. In his letters to Garran, Hunt complained that he was treated as a subordinate by his fellow officials in the Colonial Office. He wrote to Garran on 16 April:

> You can hardly conceive what extremely unimportant persons we are in the minds of the Colonial Office. Not one of these exalted gentlemen have done me the honour to leave a card, certainly Sir Campbell Bannerman did so, but to expect Mr Johnson for example, to say nothing of Bertram Cox, Lucas and the rest to have to recognise my existence would be asking a great deal too much.

Hunt was affronted by his ignominious treatment: 'I may be permitted to wait in an ante-room and even carry in any papers that may be sent for, but beyond these ennobling duties I must not infringe the sanctity of the Conference by my profane presence.'[16] Hunt was effectively denied a voice at the conference, but there was little he could do except fret and fantasise. Deakin, on the other hand, was able to speak his mind and frequently did so.

He spoke so well they thought he was beautiful

For Deakin, oratory was a key weapon in the political battle in the imperial capital for recognition of the rights and equality of Australian manhood. He had learnt about the power of speech-making in the Empire in 1887, when, as a young man of thirty, he first journeyed to London as a member of the Victorian delegation to the first Colonial Conference, called to mark Queen Victoria's Golden Jubilee. There, with 'unrestrained vigour' and 'powerful retorts', he had, by his own account, trounced the British Prime Minister, Lord Salisbury. But he had been prepared for this performance many years before—in the Deakin family home, at Melbourne Grammar School, in the university debating club and in the Eclectics Society.

Nature favoured Deakin as a public performer: he was bright, charming, tall, dark and handsome, as Gustafson had noted, and a voracious reader with a capacious memory. He was also, no doubt because of these traits, favoured by male patrons: at school, by the headmaster, Dr Bromby, and a young master, J. H. Thompson, 'a handsome athlete whom [Deakin] was not alone in worshipping', in the words of biographer La Nauze; at university, by Charles Pearson, erstwhile Professor of Modern History at King's College, London, who convened the Debating Club; in journalism, by David Syme, the influential proprietor of the *Age*, who gave him work and supported him as a twenty-two-year-old candidate for election to the Victorian Parliament, and by James Service, Victorian Premier, his patron in Parliament.[17] His biographer also found him attractive. 'His physical endowments,' La Nauze wrote with manly appraisal, 'lay in his appearance, his bearing and his voice.' Deakin's eyes 'gave vivacity to his handsome features. His voice, a light baritone in quality, had a rich timbre and a wide range. His quick gestures were adapted unconsciously to express rhetorical emphasis as he gained experience in speaking.'[18]

As a young man, Deakin was an orator in the making. His 'Notebooks' from the 1880s report his interest in studying technique and effect. From English constitutionalist Walter Bagehot, he noted the importance of constructing an argument: 'Of all the pursuits ever invented by man for separating the faculty of argument from the capacity of belief, the art of debating is probably the most effectual.' From American writer Walt Whitman, he learnt the charismatic effect of the 'right voice':

> Surely whoever speaks to me in the right voice
> Him or her I shall follow
> As the water follows the moon silently with
> Fluid steps, anywhere around the globe.[19]

Deakin had travelled part-way round the globe in 1885, on a voyage that took him across the Pacific Ocean to California to investigate irrigation, and then on to the east coast to New York and Boston, a trip whose main objective was a pilgrimage to the grave of Whitman's mentor, Ralph Waldo Emerson.

Two years later, in London, in 1887, Deakin had the opportunity for the first time to speak as a statesman on the world stage and he found exactly the 'right voice'. He wooed British statesmen and leaders of society with an 'oratorical power' that the Liberal politician Sir Charles Dilke found 'remarkable'. Dilke quickly came to the view that Deakin was 'the man of greatest promise in all Australia…a great administrator, a man of extraordinary charm and eloquence'.[20] Dilke, like many of his countrymen, seemed to find Deakin's eloquence especially surprising in 'a native born Australian', as the Parliamentary Under-Secretary, Sir William Hillier Onslow, who shared Holland's duties as president of the 1887 conference, referred to him.[21]

Shortly after Deakin's arrival, Onslow, a former governor of New Zealand, sent him an invitation to join him for dinner at Richmond Terrace, Whitehall: 'I was delighted to hear of your safe arrival. Lady Onslow is still abroad but returns this week. If you are disengaged will you give us the pleasure of your company at dinner here on Thursday 22nd at 8.15?'[22] By the time Deakin left England that year, Onslow, like so many others, had concluded that Deakin was no ordinary colonial. While other British subjects delivered themselves of platitudes and pledges of loyalty to the Mother Country, Deakin struck an entirely different note. In Deakin, they had 'a real live man'.[23]

At a dinner hosted by the Imperial Federation League, Deakin was provided with his first opportunity to speak on political matters in London, when his colleague Sir James Lorimer was unable to attend. 'I ought to have been better prepared & to have done better,' he wrote home, 'but it was not altogether a failure. Some thought it the speech of the evening.'[24] The *Age* report—from its anonymous correspondent (was it Deakin himself?)—echoed this judgment: '[I]n the unanimous opinion of all present [it was] the speech of the evening.' His theme was the imperative of Australian independence: the colonies 'were too proud to remain in such a dependent condition when they had been so amply endowed with the privileges of self-government'. The Duke of Cambridge responded with 'a warm eulogium on the ability of Mr Deakin's speech'.[25]

Deakin spoke so well they thought he was beautiful. English public men adored him and showered him with invitations to their city clubs and country houses. Initially, he had used letters of introduction from Charles Pearson (to Dilke and M. E. Grant Duff, for example), but as word spread about this charming and challenging young man from the Antipodes, so the requests for his company and literary contributions to the London magazines poured in. The editor of the *Fortnightly Review* 'seemed to take greatly to me & strongly presses me to write him a colonial article…*Murray's Magazine* has pressed the same thing but I declined all'.[26]

One of the many leading public men to whom he was introduced in London was Sir George Trevelyan, historian and politician and biographer of Macaulay. They met at a lunch hosted by Lady Holland, Trevelyan's sister, Macaulay's niece and the wife of Sir Henry Holland, the Colonial Secretary. At lunch, Deakin was seated next to Trevelyan: 'very cultured, clever & chatty,' he noted dismissively in his letter home. Though Deakin judged Trevelyan to be 'a light weight all round', he nevertheless noted down Trevelyan's remark that 'this was the age in which one fine art, that of speaking, has been brought to perfection'.[27] Deakin knew that it was an art in which he excelled.

His eloquence particularly impressed the Colonial Secretary. 'Sir H. Holland was pleased,' Deakin wrote to his sister, '& Service says he has fallen in love with me as other elders have done. I cannot see this.'[28] But he came round to this view, a few days later, when after conference discussions on the subject of defence, the Lords of the Admiralty, Sir A. Hood and Sir A. Hoskins, 'came right up to me & poured in broadsides because of my independent views of Victoria'. But then, 'Sir H. Holland came to the rescue—& Service declares his personal attentions are most marked to me—In a pleasant interview I had with him he expressed his regret at my declining the KCMG & wished me to take it. I think I have made a friend in him.'[29]

Holland was considered a proficient speaker, but not an orator, an important distinction: '[I]n style and tone he is the model of a permanent official; and, though not an orator, is quite capable of making a clear and business like statement on any point of procedure or policy.'[30] Deakin's assessments of men usually looked to their speaking capacity, which he interpreted as a reliable indicator of their manhood more generally. William Gladstone was 'strong in passion & strong in natural gifts of an oratorical order, strong in a kind of culture which fed & elevated his oratory'.[31] Poor speakers were explained as effeminate or feeble. Arthur Balfour, the young Secretary for Ireland, was 'foppish unprepossessing in appearance, egotistical and without oratorical grace'. On a visit to the House of Commons, Deakin 'heard no speaking of special order. Two feeble old Liberals moved an amendment & an aldermanish but capable conservative made a good and solid reply.'[32]

Just as a capacity for oratory was a marker of powerful manhood, so it rendered women 'masculine', as Deakin described Annie Besant, whom he heard speak in Edinburgh over Easter. His critical analysis of her performance judged it according to a number of masculinist criteria. She had an

> intense voice good & ringing not soft or winning—never smiles & her humour what there is of it grim & not pleasant—her accent quite educated & cultured. Her language clean & well chosen, admirably simple & strong. I should say her lecture was familiar to her, tho she glanced now & then at her notes. Her sentences were well formed & finished, not so flowing nor so theatrical nor so dominating as Mrs Britten, but concise & more sinewy in some respects the best speaker among women I have heard. But just as her audience was old, so was the treatment of her speech.

Deakin thought her arguments 'clean in expression & utterance, but by no means original in any single thought, utterly barren of constructive ideas'.[33] To the extent that she was a good speaker—indeed, the best speaker among women he had heard—she sacrificed her femininity. Oratory was a masculine performance; womanliness demanded a soft voice and winning smiles.

The British Prime Minister trounced

It was in the closed session of the 1887 conference, after Holland's business-like opening address, that Deakin engaged in his legendary confrontation with the Prime Minister, Lord Salisbury, who was recognised as 'a great master of expression' and hence a worthy opponent. The issue was the future of the New Hebrides. Salisbury told the colonials that they had no business interfering with imperial policy in the Pacific Ocean and that France had made a good offer in promising to cease sending convicts to New Caledonia, in return for recognition of its sovereignty over the islands. The colonial premiers responded with due deference and 'whispering humbleness', but Deakin would have none of it. Of his speech that day, the *Age* reported: 'The remarks of the junior delegate from Victoria were brightly delivered and bolder in tone than those of the preceding speakers, and enabled him for the second time since his arrival to carry off the palm of oratory from all his compeers.'[34] In a letter to his family, Deakin wrote:

> Then [the turn for the colonial representatives came] & one vied with another in congratulations felicitations & glorifications—I took quite a different key & gave them the Victorian view with some energy—It satisfied my friends & won me some praise & some declared it the speech of the day—I might and ought to have done much better.[35]

This encounter between Lord Salisbury and the junior delegate from Victoria would become central to Deakin's later account of the 'federal story', a founding moment in his history of Federation, an account in which he would transform an experience of imperial domination into a fantasy of powerful masculine triumph. Of the British Prime Minister, he wrote: 'His tone breathed the aristocratic condescension of a Minister addressing a deputation of visitors from the antipodes whom it became his duty to instruct in current foreign politics for their own sakes.' His Australian colleagues, he wrote, responded 'with bated breath and whispering humbleness, apologising for the strong feeling which had been expressed in the colonies'.[36] Then Deakin rose to his feet and, according to his own third-person account: 'He broke quite new ground not only with unrestrained vigour and enthusiasm on the general question as his colleagues had before him, but because he did so in a more spirited manner, challenging Lord Salisbury's arguments one by one and mercilessly analysing the inconsistencies of his speech.'[37]

Rising to revolutionary rhetoric, Deakin

went on to declare in an impassioned manner that the people of Victoria would never consent to any cession of the islands on any terms and that the Australian born who had made this question their own would forever resent the humiliation of a surrender which would immensely weaken their confidence in an Empire to which hitherto they had been proud to belong.[38]

The effect of such a bold statement, he reported, was 'electrical'. 'Lord Salisbury several times stared at the speaker, as well he might…[in] considerable amazement at his plain speaking and in some discomfort at the stern debating retorts to his inharmonious contentions.'[39] The British Ambassador in Paris was promptly informed that nothing was to be ceded to the French. Deakin was undoubtedly pleased with his performance, but the lesson he drew from his experience at the Colonial Conference was that had the colonies spoken in unison, with one voice, their achievements would have been all the greater. In union lay power. Small wonder that the next year Royce concluded that Deakin liked power and that separation from Britain was inevitable.[40]

In conclusion: only talk

Deakin's experience of colonial subjection was, by his own account, demeaning and humiliating, yet he managed to transform it, in memory, into a story of potent triumph. Yet his victory over the British Prime Minister, like Atlee Hunt's impressive speech-making, was largely a consoling fantasy. Even with the achievement of federal union, Australia remained in a condition of colonial dependency, subject to British power, still complaining about French (and German and Japanese) intentions in the Pacific for years to come.

Hence the anxiety that accompanied Deakin's attendance at the conference in 1907. Hunt noted that Deakin was plagued by over-worry and indigestion.[41] Deakin referred to the conference in 1907 as an imperial conference, a gathering he fantasised as a meeting between equal autonomous governments, the same meaning he would give to 'Imperial Federation'. He arrived in London determined to remove all power from the Colonial Office, but had to make do with a promise to establish a new Dominions Office, headed by veteran Colonial Office man Sir Charles Lucas (one of the officials who had ignored Secretary Hunt).[42] Deakin's capacity for oratory, in his view, redeemed Australian manhood, but in the event it also served to consolidate his country's dependency.

While Deakin raged inwardly at Australia's humiliation—at being denied the right to contribute to defence and foreign policy, at Australia being treated as an object of negotiation between the great powers—he took consolation in his oratorical powers. Deakin had mastered the art of oratory and oratory had mastered him. Oratory offered Deakin (and in his dreams, Secretary Hunt) the taste of masculine power. But Deakin also knew the limits of 'talk'. In his first letter home from England on 31 March 1887, he announced: 'I have made my debut in London society & am prepared to give judgment upon it at once.' He had attended an 'At Home' of Countess Stanhope.

She was very glad to see me she said, tho' I doubt if she knew where I came from & probably said the same thing to a couple of hundred others. Somebody else was announced right on my heels & so I passed…The dinners are just as our dinners, people dress better & more people talk well. Some brilliantly, but it is only talk & therefore worth nothing.[43]

Oratory was an art and a gift, and it lifted some men above others, but it was, after all, just talk.

In oratory, Deakin sought to redeem Australian manhood, and his own speech-making undoubtedly played a significant role in the achievement of Federation, yet his country remained in a position of colonial dependency, subservient in important matters to the Colonial Office. The strain generated by the tension between that dependency and his republican desire for manly independence took its psychic toll. Biographer La Nauze noted that after the 1907 conference, something was damaged in Deakin and could not be repaired. W. M. Hughes, looking at Deakin across the benches of the House of Representatives on his return, concluded that Deakin had had a nervous breakdown.[44]

ENDNOTES

[1] Arthur Atlee Hunt to Bob Garran, 16 April 1907, Hunt papers 52/767, National Library of Australia.

[2] Hunt to Garran, 25 April 1907, Hunt papers 52/772A.

[3] Ibid.

[4] Ibid.

[5] Ibid.

[6] Hunt to Garran, 25 April 1907, Hunt papers 52/772B.

[7] La Nauze, J. A. 1965, *Alfred Deakin A Biography*, Vol. 2, Melbourne University Press, Melbourne, p. 500.

[8] Lake, Marilyn 2007, '"The Brightness of Eyes and Quiet Assurance Which Seem to Say American": Alfred Deakin's Identification with Republican Manhood', *Australian Historical Studies*, Vol. 38, No. 129, pp. 32-51.

[9] Ibid.

[10] *Scribner's Magazine*, IX, No. 1 (January 1891), p. 78.

[11] Sir Francis Hopwood, note, 4 April 1908, CO418/60/105.

[12] Zadel Gustafson to Alfred Deakin, 6 March 1890, Deakin papers 1540/11/4, National Library of Australia.

[13] Ibid.

[14] Ibid.

[15] Hunt to Garran, 18 April 1907, 52/769.

[16] Hunt to Garran, 16 April 1907, 52/767C.

[17] La Nauze, *Alfred Deakin*, Vol. 1, p. 18.

[18] Ibid., p. 24.

[19] Notebooks, Deakin papers 1540/3/11.

[20] La Nauze, *Alfred Deakin*, Vol. 1, p. 91.

[21] Ibid., p. 94.

[22] Onslow to Deakin, no date, Deakin papers 1540/9/344.

[23] La Nauze, *Alfred Deakin*, Vol. 1, p. 94.

[24] Deakin travel diary, Deakin papers 1540/2/39/.

[25] News clipping, undated report by special correspondent (Deakin himself perhaps?), Deakin papers 1540/9/478.

[26] Travel diary, Deakin papers 1540/2/39/.

[27] Ibid.

[28] Ibid.

[29] Ibid.

[30] *Age*, 14 May 1887.

[31] Travel diary, Deakin papers 1540/2/39.

[32] Ibid.

[33] Ibid.

[34] *Age*, 14 May 1887, Deakin papers 1540/9/474.

[35] Deakin's travel diary, 1887, 2/2/39.

[36] Deakin, Alfred 1963, *The Federal Story*, edited and introduced by J. A. La Nauze, Melbourne University Press, Melbourne, p. 21.

[37] Ibid., p. 22.

[38] Ibid., p. 23.

[39] Ibid., p. 23; La Nauze makes the point that there were other slightly different, less heroic, versions of the encounter than Deakin's, but what interests me is precisely Deakin's version of it. See La Nauze, *Deakin*, Vol. 1, pp. 96–7.

[40] Royce, Josiah 'Reflections after a Wandering Life in Australasia', *Atlantic Monthly*, Vol. LXIII, p. 827.

[41] Hunt to Garran, 16 April 1907, 52/767.

[42] Ibid.

[43] Deakin's travel diary, 1887, 2/2/39.

[44] LaNauze *Deakin* Vol. 2, p. 514; Hughes to Jebb quoted in LF Fitzhardinge, 1964, *William Morris Hughes A Political Biography* vol. 1, Angus and Robertson, Sydney, p. 191.

5. Hunting the Wild Reciter: Elocution and the Art of Recitation

Peter Kirkpatrick

Despite the endurance of bush poetry festivals and an inner-city fashion for contemporary spoken word and performance poetry, these days the recitation of verse in Australia—as in other Western anglophone nations—remains a minority taste. Early last century, things were very different. In the 1920s, Australians were reciting verse all over the place; so much so that the well-known *Bulletin* humorist 'Kodak' O'Ferrall lampooned them:

> Way out in the suburbs howls the wild Reciter,
> Storming like a general, bragging like a blighter;
> He would shame hyenas slinking in their dens
> As he roars at peaceful folk whose joy is keeping hens.
> 'How We Beat the Favourite', 'Lasca', 'Gunga Din',
> There they sit and tremble as he rubs it in.
> When he's done they *thank* him! Never do they rise,
> Tie his hands and gag him as he rolls his eyes,
> Bag his head and bear him swiftly through the night.
> *That's* the only remedy for villains who recite.[1]

Here, from the modern, metropolitan perspective of the professional writer, the reciter is an outlandish form of *Homo suburbiensis*. An amateur with half-baked literary taste, his choice of party pieces is quaintly Victorian, comprising narrative ballads from Adam Lindsay Gordon to Rudyard Kipling, and extending to *Lasca* by the Englishman Frank Desprez, in which a Texan cowboy laments the death of his Mexican sweetheart in a cattle stampede. And yet, despite his vulgarity, the reciter's manic energy makes him a force to be reckoned with. It seems only physical violence can put an end to his kind.

This chapter traces the rise and fall of the wild reciter in Australia. In doing so, it takes recitation to refer to the memorisation and public performance of printed verse not written by the reciter, whose repertoire sometimes incorporated the work of canonical poets, but which tended to focus on more popular authors (including stage entertainers), who are now largely forgotten. Reciters might also have included a range of prose highlights and dramatic character sketches in their recitals, but it's evident that rhyming verse in a variety of traditional metres was favoured, and this will be the focus here.

Analysing the historical sources of recitation offers not merely a wistful glimpse into one of the lost arts of everyday life in Australia. It also sheds light on some of the ways in which poetry was mobilised by modern popular culture and the new technologies that served it. What follows revisits the broader cultural context in which recitation flourished, outlining three factors that led to its enormous popularity: the growth and spread of elocution, the institutionalisation of recitation in schools, and its professionalisation as a performance art. As the last two are substantially a consequence of the first, elocution will be thematic throughout this account. Indeed, the fate of recitation was tied inextricably to that of elocution—as we shall see by way of conclusion.

Elocution began in the eighteenth century as a means of training lawyers and churchmen in oratory, but by the end of the nineteenth century, it had broadened into a middle-class movement concerned to foster proper and 'natural'—that is to say, middle-class—speech habits. In the words of Dwight Conquergood,

> Elocution was designed to recuperate the vitality of the spoken word from rural and rough working-class contexts by regulating and refining its 'performative excess' through principles, science, systematic study, standards of taste and criticism…Ambivalently related to orality, elocution sought to tap the power of popular speech but curb its unruly embodiments and refine its coarse and uncouth features.[2]

In a world without electronic means of amplification, elocution also taught voice projection for all forms of public speaking, professional and amateur, including that for the theatre. The great influence of elocutionary practices throughout the nineteenth century on public entertainments such as lectures, recitals and music hall, in conjunction with an ever-expanding market for ballads and topical verse in newspapers, made the speaking of poetry very popular indeed.

The poetry generation

In her splendid study of *The Victorian Popular Ballad*, J. S. Bratton describes the way in which Romantic experiments with the literary ballad interacted with older traditions such as the broadside to produce a new kind of popular verse. An early favourite was Lord Macaulay's *Lays of Ancient Rome* (1842), which set a fashion for historical ballads across the Empire that lasted well into the twentieth century, and which became institutionalised through recitation in imperial education systems. Macaulay's *Lays* were set for first-year study in the New South Wales secondary syllabus between 1911 and 1944[3] and clearly had a long afterlife. In the late 1960s, I can recall coming across an extract from *Horatius* in an early high-school anthology called *Poems of Spirit and Action* [4] —an old-fashioned collection that also included naval ballads by Henry Newbolt, cowboy poems by Bret Harte and William Rose Benét, as well as *The Sick Stockrider* by Adam Lindsay Gordon and *In The Droving Days* by Banjo Paterson.

The success of the bush ballad in the Australian context needs to be scrutinised in the light of this wider, imperial taste for literary balladry. Because of its status in white Australian folklore, radical nationalists such as Russel Ward saw the bush ballad as a uniquely demotic form, a function of the vernacular realism sponsored by the *Bulletin*. Yet Richard White has shown how the *Bulletin*'s image of the heroic frontiersman *à la The Man from Snowy River* was in fact a common literary trope across the British Empire and the United States in the late nineteenth century (*Lasca* is a perfect case in point). He compares the topos of the city versus the bush in Henry Lawson's work, for instance, with the celebration of colony over metropolis in Kipling: 'It must also be remembered that the bush-worker was an integral part of empire and, when he was ennobled as "the Bushman" and his capacity for drunkenness and blasphemy forgotten, contributed much to imperial ideology.'[5] Read within this frame, many bush ballads appear less a folkloric expression of Australian national consciousness than local variations of a hegemonic literature of colonial adventure.

The genre of the heroic ballad was so widely popular that, across all class levels, schools encouraged patriotic sentiment through the declamation of prominent examples.[6] As Terry Eagleton has pointed out, it's hardly coincidental that Henry Newbolt, bard of such Boys' Own classics as *Drake's Drum* and *Vitai Lampada*, should also have written the highly influential 1921 British government report *The Teaching of English in England*.[7] Bratton notes:

> It was this appropriation to schoolroom use…which converted the heroic
> historical ballad into the staple popular poetry of the middle classes, and
> imposed upon it the peculiarity of becoming as much oral as written
> poetry. Children were set to learn these instructive and inspiring pieces;
> and what is learned must be recited.[8]

Similarly, in the American context, Joan Shelley Rubin has described patriotism as one of three 'prevailing purposes' in school recitation, the other two being 'the cultivation of moral sense and the desire to equip the young with memorised works that could provide "comfort, guidance, and sympathy" throughout life'.[9]

Elocutionary belief in the virtue of 'good' speech underpinned recitation in schools. As a popular form of speech training or 'correction', it was important in reshaping individuals, facilitating class mobility by disguising working or lower-middle class or provincial accents—including, of course, Australian accents. The extension of elocution's social function was related directly to the expansion of public education, and consequent concerns about the cultural impact of mass literacy that went hand-in-hand with the rise of English as a subject for schools and an academic discipline. The recitation of verse was seen as a useful skill for 'raising' and standardising pronunciation and thus, supposedly, helping to overcome social division.[10]

In their 1992 oral history of Australian reading practices, Martyn Lyons and Lucy Taksa describe the generation that grew up before and during World War I as 'The Poetry Generation', and their interviewees (61 of them, all aged over seventy, born between 1886 and 1917) reported encountering the work of the following poets:[11]

Writers	Mentions
Henry Lawson	32
William Shakespeare	31
Banjo Paterson	29
Henry Longfellow	21
C. J. Dennis	20
William Wordsworth	18
Alfred Lord Tennyson	15
Adam Lindsay Gordon	14
Rudyard Kipling	10
Mary Gilmore	10
John Milton	10

Though Lyons and Taksa are interested in reading rather than recitation, their figures make no clear distinction between these different modes of consumption. As they point out: 'The poetry in the imaginary library of the Australian reader, which we have tried to reassemble here, did not sit silently on the page. It was memorised and recited.'[12] The range of poets recalled by their interviewees represents a span from the low and local (Lawson, Paterson, Dennis), through a middle rank of popular Victorians (Longfellow, Tennyson), to the high and canonical (Shakespeare, Wordsworth, Milton), emphasising the extent to which poetic taste was informed by a school education system that fashioned Australian identity in terms of imperial citizenship. In this list, only three Australian poets and Kipling—and then largely on the basis of their longevity—might be considered twentieth-century writers, and none are modernists.

Rubin's account of recitation in American schools between the wars similarly notes a preference for nineteenth-century material: 'The prescribed texts in 1935 were frequently the same ones tested on college entrance examinations in 1890.'[13] She makes the important point that such old-fashioned and occasionally eclectic selections serve to distinguish histories of reading from more conventional literary histories, which privilege innovation: 'Thus the recitation provides a graphic illustration of the discrepancy between textual production and consumption at any given time.'[14]

Pieces selected for reciter books and elocution manuals also offer evidence of this conservative, by now almost customary, form of textual circulation. The eponymous hero of V. S. Naipaul's Trinidadian novel, *A House for Mr Biswas*, misappropriates a copy of the much-reprinted *Bell's Standard Elocutionist* as a schoolboy, and the book stays with him throughout his life, as 'his favourite reading'[15] and as the repository of a cultural capital that, however outmoded and irrelevant to his own context, he can never accrue. Biswas was taken out of school just as he had started to memorise *Bingen on the Rhine* by Lady Caroline

Norton—an English poem about a dying French legionnaire in Algiers who pines for his hometown in Germany—and it is up to his educated son Anand (a version of the young Naipaul himself) to achieve what was denied his father:

> He increased the pathos in his voice, spoke more slowly and exaggerated his gestures. With both hands on his left breast he acted out the last words of the dying legionnaire.

> *'Tell her the last night of my life, for ere this moon be risen,*
> *My body will be out of pain, my soul be out of prison.'* [16]

Anand's touching performance takes place in the 1940s, when the cultural cringe it represents was still common across the British Empire.

Hugh Anderson's account of reciter books published in Australia between the 1860s and the 1940s reveals that most comprised British and American pieces with a few Australian poems thrown in for local colour. [17] Among the more enthusiastic anthologists was W. T. Pyke, manager of Cole's Book Arcade in Melbourne, who edited three collections: *The Applause Reciter* (1898), first published in London, which had no Australian content; *The Coo-ee Reciter* (1904), with two-thirds Australian material; and *The Australian Favourite Reciter* (1907), with one-third Australian material. The last was republished as late as 1947, 14 years after Pyke's death, with its 1907 preface and spectral invitation for more contributions intact, and still prophesying 'two or three more books on similar lines'. [18]

The great exception to this transnational pattern of choice is *The Bulletin Reciter*, selected by A. G. Stephens and first published in 1901, which garnered only Australian verse that had appeared in that journal in the previous two decades. It was the most popular of the locally produced reciters and went through at least 14 editions, [19] the final edition appearing in 1940 and boasting on its cover 'Over 250,000 copies sold'. The original *Reciter* was replaced in 1920 by a new, updated collection edited by Bertram Stevens, called *The Bulletin Book of Humorous Verses and Recitations*, which went through at least three editions. In 1933, however, the old *Reciter* was reinstituted with 'enlarged' contents, [20] its return highlighting not only the conservatism of popular taste but that recitation was already moribund.

Clearly the success of the reciter books is explained only partly by the promotion of recitation in schools. As already noted, outside the classroom, recitation was also practised widely as a form of entertainment—not only by amateurs, but by professionals from a range of cultural milieux, from local variety performers to world-renowned Shakespearean actors. Their performances are the most visible factor in the popularity of spoken verse, yet, as much as the schoolchild stammering through his or her set piece, they also took place in a world highly attuned to accent and the auditory discipline of elocution.

Resuscitating an immoral piece of poultry

Poetry was genuinely popular during the Victorian era in a way that it has never been, before or since. As Bratton observes, social and technological innovations 'from the growth of the industrial towns to the diversification of the periodical press' changed the audiences for poetry and their literary taste.[21] Les Murray has thus praised the verve and variety of nineteenth-century newspaper poems in Australia, and the way in which popular journals in those days worked to cross-fertilise the high end of verbal art with the low. The Victorian period represented 'the narrow-columned middle ground' of poetry, before the 'social-divide model of Australian verse'[22] had started to appear, after which literary—and, by implication, modernist—elites overruled the mass of poetry lovers.

It must be remembered that the taste for recitation arose alongside that for public reading, itself the product of an elocutionary culture that sought to regulate the voice as a sign of gentility. Philip Collins has traced the fashion for reading aloud in Britain to the establishment of Mechanics' Institutes, where, from the 1840s, 'readings and other forms of respectable light entertainment came to supplement (later rather to displace)' educational lectures.[23] According to Collins, 'The common readers' enjoyment of some of the best authors of the age—most notably, of Tennyson and Dickens—must have been reinforced and extended by their hearing of so many of their pieces in recitals, public and private.' As a result, 'many people met contemporary literature as a group or communal, rather than an individual experience'.[24]

The relationship between poetry and song during this period is also important. As Murray observed: '[A]s in previous ages, the sharp distinction we make between poetry and song would have seemed strange.'[25] The Victorian stage was overwhelmingly musical in character, from opera and operetta down to melodrama, fairy ballets and pantomime, but its supreme popular expression was the music hall, in which 'all branches of popular entertainment came to be absorbed'.[26] With its comic monologues and sentimental ballads, music hall was a powerful influence on the popularity of recitation, and artists such as Bransby Williams, renowned for his rendition of J. Milton Hayes' *The Green Eye of the Little Yellow God*, became famous. In this context, a clear distinction between recitation and character sketches and songs is often difficult, especially as songs performed by actors were sometimes delivered in a kind of *Sprechgesang*. In Australia, music halls, as such, had a relatively short history, and were overtaken in the late 1860s by variety theatres, which appealed to a broader range of classes, and which in turn laid the foundation for the later Tivoli vaudeville circuit, though the style remained derivative of English music hall.[27]

Accordingly, the twentieth-century vaudeville comic Mo (Roy Rene) had a routine in which he attempted to recite *The Green Eye of the Little Yellow God*, only to be interrupted constantly by stooges planted in the audience. Driven to distraction, Mo would splutter, 'Oh this is lovely! This is beautiful! A gentleman and a scholar can't get up to resuscitate an immoral piece of poultry without being got at!'[28] But even Mo could take himself quite seriously as a reciter. On stage and later on radio, he would occasionally come out with *Life is a Very Funny Proposition After All*, a mangled version of a piece of homespun wisdom by the American popular entertainer George M. Cohan. As Alexander Macdonald recalled:

> He'd long forgotten most of the words of this piteous lament, but that was no obstacle. He filled in with a few of his own. And the finished product emerged as quite a remarkable length of double-talk:

> *Life is a very funny proposition after all,*
> *Imagination, hypocrisy and gall,*
> *With not much to say, and two meals a day,*
> *And when you're broke you're always in the way.*
> *And waiting and waiting, and still no curtain call…*
> *Life is a very funny proposition after all.*

> After putting this across he'd come back stage flushed with triumph. 'I had the mugs weeping,' he'd say, 'Bloody well weeping.'

'They were weeping, all right,' Macdonald added, 'weeping with laughter.'[29] Yet Fred Parsons, Macdonald's co-writer on the Colgate–Palmolive Radio Unit, who tells much the same story, stated that the audience 'would hang on every word' and give the comic 'a terrific hand'.[30]

These stories of Mo's recitations recall the 1940s, when he was a national celebrity, and they mark not only the twilight of public recitation as a performance art but two distinct attitudes to it. On the one hand, the postures and poetic discourse of the reciter provide easy objects of satire. On the other hand, publicly performed verse—however trite—was still able to move an audience. Mo could, with some justification, believe that his rendering of *Life is a Very Funny Proposition After All* left the mugs in tears.

Outside of music hall and vaudeville, recitation was a predominantly genteel activity, a perceived social accomplishment in the Victorians' endless quest for self-improvement. Middle-class audiences who wouldn't dream of joining the throng 'Down at the Old Bull and Bush' could hear sensational, pathetic or comic verses performed in the parlour or concert hall. As Bratton observes, '[P]erformers like Bransby Williams…were acceptable not for their personal gifts or background but because their material was common to all.' She describes a style of presentation that 'could be transposed from music hall to concert room, even to church hall or private house':

The ingredients were a strong voice, and a magic lantern: the performer stood in strong spot-light or in dramatic silhouette, and a series of slides depicting scenes in the ballad were thrown on the wall or screen behind, while the artist delivered the ballad with all the dramatic gesturing and vocal acrobatics at his or her command.[31]

As a means of showcasing elocutionary skill, recitation was also part of a professional stage actor's training.[32] Kenneth Pickering writes: 'It is easy for us to overlook the importance of voice to the nineteenth-century actor and audience...[but] elocutionary display was an accepted and integral part of theatre'; indeed, '[t]o some extent "elocution" became synonymous with acting'.[33] This involved acquiring a knowledge of prosody, as well as a facility for the semi-dramatic recital of poetry. (It should be recalled, too, that verse plays were commonly produced in the Victorian era; not just those of Shakespeare and his contemporaries, but what we would now regard as the closet dramas of poets from Byron to Tennyson.) Some of the great Victorian actors became identified with their rendition of particular poems—for example, Henry Irving with Thomas Hood's eerie confessional *The Dream of Eugene Aram*—and the widespread popularity of recitation produced a new kind of professional platform recitalist who would often use music to enhance the emotional effect.[34] In Australia, one such was Lawrence Campbell, who was seen as pre-eminent in his field,[35] and a typical recital was described in the *Theatre Magazine* in 1905:

> The principal interest in an entirely new programme will centre [on] Mr Campbell's delivery of Tennyson's well-known poem, 'The Lady of Shalott', with a piano, violin and 'cello accompaniment.
>
> Other elocutionary items will include a selection from *Macbeth*, Act IV, Scene I, characters represented—Malcolm, Macduff, and Rosse [sic]. 'The Village Priest', by Farcalladen Moor, with musical accompaniment by Miss Lilian Frost; and two humorous sketches, 'The Family Vault', by Max Adler, and the 'Two Scars', by Overton.[36]

Though a distinction was drawn between platform and stage performance, the perceived benefits of elocutionary training allowed considerable overlap. *Theatre Magazine* described veteran actor William Holman as 'one of the most popular reciters and successful teachers of elocution in Sydney', who had coached state and federal MPs—including his own son, the future Labor Premier W. A. Holman—in voice production.[37] Advertisements in the journal for drama teachers constantly link platform and stage. Holman offered lessons in 'elocution, voice culture, dramatic art', and Campbell described himself as a 'teacher of elocution and dramatic art'.[38] Actor-manager Duncan Macdougall, who set up Sydney's pioneering Playbox Theatre in the 1920s, began his career as a teacher of elocution and trained future Presbyterian ministers in public speaking at St Andrew's College at Sydney University.[39]

The solid popularity of recitation in the early 1900s makes its sudden decline by mid-century the more remarkable. After all, it had behind it the institutional support of the education system and many prominent people in public life, including some in what we would now call the entertainment industries. Encouraged by a network of teachers of elocution, and fed by an ever-expanding range of recitation books, the wild suburban reciter was in his glory. To find out why public taste turned so decidedly against him—why Kodak wanted to 'Bag his head and bear him swiftly through the night'—we need to consider some effects of the new audial landscape that arrived with the turn of a dial shortly after World War I.

Radio killed the poetry star?

What killed recitation as a popular art? In *The Oxford Companion to Australian Folklore*, Keith McKenry blames the new media technologies 'combined with normal social change' for the decline of recitation from the 1920s.[40] If this is true, it is not *simply* true, as radio arguably brought new opportunities for extending poetry as entertainment across the airwaves. The ABC's John Thompson, who himself fostered such hopes through programs such as *Quality Street*, quoted the English poet D. G. Bridson, who in 1950 believed that there was 'every reason to class the invention of the microphone with those of writing and the printing press'. By 1966, however, as Thompson ruefully remarked, 'Bridson's hopes [had] hardly been fulfilled'.[41] What had happened?

Firstly, as Les Murray observed, poetic culture began to bifurcate under the impact of modernism, with its emphasis on a less public, more intimate voice modulated in free verse. Designed to be sounded mentally rather than spoken out loud, modernist verse was tailored to an audience of individual readers, not gatherings of auditors. Fuelling this shift in attitude was the fact that elocutionary techniques did not transfer well to the electronic media. The informal kind of voice required for radio was precisely the opposite needed for histrionic declamation in large, draughty halls and theatres, and, as listeners became quickly habituated to the paradoxical intimacy of public broadcasting, the former advantages of elocutionary training became liabilities. In this process, the mnemonic and incantatory effects of traditional metres were downgraded, becoming associated with complacent and conservative social attitudes—what George Orwell, in a famous critique of Kipling, called 'good bad poetry'.[42]

Secondly, recitation was strongly associated with the artificiality of elocution, and therefore suffered by association when performance fashions turned against it. Elocution always had a problematic relationship with 'nature'. In their attempt to reform the common vices of static or unanimated oratory, the eighteenth-century pioneers of elocution had endeavoured to develop a universal grammar of emotion linked to bodily gesture: '[N]ature has given to every emotion of the mind its proper outward expression,' wrote James Burgh, author of *The Art of Speaking* (1762).[43] Not surprisingly, as the elocutionary movement burgeoned during the nineteenth century in a host of manuals illustrating recommended poses and hand movements, this simple idea became rigidly codified. As Frederick William Haberman has written: 'The inevitable result of this conception of natural determinism in emotional expression is the degeneration of the philosophy of delivery into a study of symbolical expression, or formalized action, of regulated postures, of designated maneuvers [sic].'[44] In a version of *Bell's Standard Elocutionist* republished as late as 1938, the reader is told 'that gesture must appear spontaneous, must seem to rise, as a result of unstudied impulse, from the words which are being spoken'.[45] Yet two pages later, under the heading 'Full Gesture', we are told:

> Every accentual action should be preceded by a prepatory movement in the opposite direction to where the movement will end, the gesture needing somewhere to move *from*.

> As a general principle, either the upper side (the thumb), or the back of the hand, should be turned in the direction of the gesture in upward motions; and the lower side, or the palm of the hand, in downward motions.

> The feet should preserve a uniform angle of separation in every motion, the toes being turned outwards to an angle of not more than 75 degrees.[46]

So much for spontaneity! Haberman argues that, as a product of Enlightenment scientism, elocution was always mechanistic in its approach, even though its originators saw themselves as merely following natural precepts: 'Implicit in the minds of the elocutionists was the sense of a mechanical order in nature. Thus they could claim that their rules and their systems represented the order that is found in nature; they were "nature still, but nature methodized".'[47]

Even before radio, in the theatre, the advent of naturalism in the late nineteenth century started to give the histrionics of elocutionary training a bad name—a situation, as Kenneth Pickering notes, 'greatly complicated by the large numbers of amateur performers trying to teach themselves the socially desirable skill of recitation'.[48] Writing on 'The elocutionist and the actor' in *Theatre Magazine* in 1913, Lawrence Campbell commented:

> If one thing more than another has helped to bring the elocutionary profession into disrepute it is the failure on the part of the majority of interpreters, both professional and amateur, to realise the limitations of their art—a failure which has lead [sic] to an excessive and exaggerated

form of action and impersonation as artificial and unconvincing as it is strenuous and inartistic.[49]

To take just one example, for 20 years until his death in 1950, an eccentric known only as The Tiger, or the Tangalooma Tiger, and describing himself as 'a full-blooded elocutionist', drew substantial crowds to his recitations of Henry Lawson and Shelley at the Sydney Domain.[50]

Recitation and elocution were at last parting company, and if elocution itself was changing it was a little too late to save its name. Mark Morrisson describes the emergence of a 'new elocution' in Britain at the turn of the twentieth century that valued restraint and a lack of theatrical effects, emphasising the reciter's role in bringing forth the latent 'pure voice' of the poem itself. He quotes Lady Margaret Sackville, whose inaugural address to the Poetry Society in 1912 advised that '[t]here should be no striving from outside to produce a definite effect—the soul of the interpreter should be so possessed by the poem that it follows it instinctively in every modulation and inflection as easily as water flowing between winding banks'.[51] Though the new elocutionists were literary conservatives, Morrisson argues that their stress on unadorned performance—the purely oral over the visual and dramatic—was a powerful influence on early modernist poetry's desire to break free of Victorian mannerisms and, in T. S. Eliot's words, 'purify the dialect of the tribe'.

In the Australian context, these ideas are implicit in the comments of one elocutionist, who told readers of the *Age* in 1938: 'The word "elocution" is associated in the minds of most people with recitations, or as some artificial manner of speech. This is easily understood, because in schools and academies reciting is the only medium through which elocution is taught.'[52]

True elocution, the writer believed, was simply the art of speaking clearly so as to fully articulate the sound of the words. The year before, in *Australian Musical News*, an 'Examiner in Elocution' similarly insisted: 'Audibility, good vocal tone, rhythmic flow, naturalness, and sincerity, coupled with understanding and feeling, are, in short, the essentials of good verse speaking, and the lack of these qualities can never be compensated for by artificial expression or the theatrical methods dear to the elocutionists of the past.'[53]

In her history of early Australian radio, Lesley Johnson has described the evolving style of presentation from the 1920s, when 'advertisements for new programmes occasionally mentioned that the announcer was a "fine elocutionist"', to the 1930s, when radio 'personalities', noted for their intimate, friendly manner, started to come into their own.[54] In the 1950s, when the career of radio superstar and sometime poet John Laws was just beginning, one Western Victorian reciter gave up performing because he felt that he and his fellows were now regarded as 'bits of fools'.[55] Radio thus helped kill the wild reciter, not simply by offering alternative entertainment, but by finally dispensing with the last vestiges of an elocutionary ethos that had sustained the art of recitation for more than a century. Elocution flourished on the stage and in the public readings and recitals of early industrial society, but the new audiences so formed would become the basis of a modern mass culture that has so far proved mostly unfriendly to poetry.

ENDNOTES

[1] Kodak (Ernest O'Ferrall) 1928, *Odd Jobs*, Art in Australia, Sydney, n.p.

[2] Conquergood, Dwight 2000, 'Rethinking elocution: the trope of the talking book and other figures of speech', *Text and Performance Quarterly*, Vol. 20, No. 4, pp. 326–7.

[3] Watson, Ken 1996, 'English Teaching in Historical Perspective: Case Study of an Australian State', in Ken Watson (ed.), *English Teaching in Perspective in the Context of the 1990s*, Third Edition, St Clair, Sydney, pp. 28–9.

[4] See Smyth, W. M. (ed.) 1957, *Poems of Spirit and Action*, Edward Arnold, London, pp. 126–33.

[5] White, Richard 1981, *Inventing Australia: Images and Identity 1688–1980*, George Allen and Unwin, Sydney, p. 103.

[6] See Bratton, J. S. 1975, 'The Heroic Ballads', *The Victorian Popular Ballad*, Macmillan, London and Basingstoke.

[7] Eagleton, Terry 1983, *Literary Theory: An Introduction*, Basil Blackwell, Oxford, p. 28.

[8] Bratton, *The Victorian Popular Ballad*, p. 49.

[9] Rubin, Joan Shelley 1997, '"They flash upon that inward eye": poetry recitation and American readers', *Proceedings of the American Antiquarian Society*, American Antiquarian Society, Worcester, Mass., p. 275.

[10] See Muddlestone, Lynda 1995, 'Educating Accents', *'Talking Proper': The Rise of Accent as Social Symbol*, Oxford University Press, Oxford, especially pp. 308–9.

[11] Lyons, Martyn and Taksa, Lucy 1992, *Australian Readers Remember: An Oral History of Reading 1890–1930*, Oxford University Press, Melbourne, pp. 59–60.

[12] Ibid., p. 68.

[13] Rubin, '"They flash upon that inward eye"', p. 274.

[14] Ibid., p. 275. Also see Rubin, Joan Shelley 1998, 'Listen, My Children: Modes and Functions of Poetry Reading in American Schools, 1880–1950', in Karen Halttunen and Lewis Perry (eds), *Moral Problems in American Life: New Perspectives on Cultural History*, Cornell University Press, Ithaca, NY, p. 261.

[15] Naipaul, V. S. 1969, *A House for Mr Biswas*, Penguin, Harmondsworth, p. 161.

[16] Ibid., p. 415. Recitation of this poem also features, more romantically, in the Canadian novel *Anne of Green Gables*.

[17] Anderson, Hugh 2001, 'That face on the bar-room floor, or, Where have all the reciters gone?', *Australian Folklore*, No. 16, pp. 190–3.

[18] Pyke, W. T. 1947, 'Preface', *The Australian Favourite Reciter*, J. Pollard, Melbourne, n.p.

[19] Anderson, 'That face on the bar-room floor', p. 191.

[20] [Stephens, A. G. (ed.)] 1933, *The Bulletin Reciter: Collection of Verses for Recitation: From the* Bulletin: *Enlarged Edition*, NSW Bookstall Co., Sydney. The contents were expanded by the addition of a mere four poems by W. T. Goodge. By this time, *The Bulletin Book of Humorous Verses and Recitations* was now advertised as *The Bulletin Reciter No. 2*, endpapers.

[21] Bratton, *The Victorian Popular Ballad*, p. 12.

[22] Murray, Les 1992, 'The Narrow-Columned Middle Ground', *The Paperbark Tree: Selected Prose*, Carcanet, Manchester, p. 240.

[23] Collins, Philip 1972, *Reading Aloud: A Victorian Métier*, The Tennyson Society, Lincoln, p. 11.

[24] Ibid., p. 27.

[25] Murray, 'The Narrow-Columned Middle Ground', p. 233.

[26] Bratton, *The Victorian Popular Ballad*, p. 27.

27 Waterhouse, Richard 1995, 'Music Halls', in Philip Parsons (ed.), *Companion to Theatre in Australia*, Currency with Cambridge University Press, Sydney, p. 381.

28 Macdonald, Alexander 1972, *The Ukelele Player Under the Red Lamp*, Angus and Robertson, Cremorne, p. 224.

29 Ibid., pp. 229–30.

30 Parsons, Fred 1973, *A Man Called Mo*, Heinemann, Melbourne, p. 83.

31 Bratton, *The Victorian Popular Ballad*, p. 107.

32 Pickering, Kenneth 1984, 'The rise of elocution—continued', *Speech and Drama*, Vol. 33, No. 1, p. 4.

33 Pickering, Kenneth 1983, 'The rise of elocution', *Speech and Drama*, Vol. 32, No. 2, p. 9.

34 Pickering, 'The rise of elocution—continued', pp. 7–8.

35 He featured on the cover of *Theatre Magazine*, 1 July 1919, as 'Australia's most distinguished platform entertainer'.

36 'Mr Lawrence Campbell', *Theatre Magazine*, 1 April 1905, p. 10.

37 'Mr William Holman', *Theatre Magazine*, 1 April 1905, p. 8.

38 *Theatre Magazine*, 1 November 1911, p. 14.

39 Kirkpatrick, Peter 2000, 'Macdougall, Augusta ("Pakie") and Duncan', in John Ritchie (ed.), *Australian Dictionary of Biography*, Vol. 15, Melbourne University Press, Carlton South, p. 199.

40 McKenry, Keith 1993, 'Folk Poetry and Recitation', in Gwenda Bede Davey and Graham Seal (eds), *The Oxford Companion to Australian Folklore*, Oxford University Press, Melbourne, p. 152.

41 Thompson, John 1966, 'Broadcasting and Australian Literature', in Clement Semmler and Derek Whitelock (eds), *Literary Australia*, F. W. Cheshire, Melbourne, p. 97.

42 Orwell, George 1970, 'Rudyard Kipling', in Sonia Orwell and Ian Angus (eds), *The Collected Essays, Journalism and Letters of George Orwell: Volume II: My Country Right or Left 1940–1943*, Penguin, Harmondsworth, pp. 226–8.

43 Cited in Haberman, Frederick William 1947, The elocutionary movement in England, 1750–1850, PhD, Cornell University, p. 113.

44 Ibid., p. 129.

45 [Bell, Alexander Melville] 1938, *The Standard Elocutionist: A Complete Book on the Science and Art of Easy, Clear, Effective and Expressive Speech, Compiled, with Revisions and Additions, from Bell's Standard Elocutionist*, English Universities Press, London, p. 43.

46 Ibid., p. 45.

47 Haberman, The elocutionary movement in England, 1750–1850, p. 50.

48 Pickering, 'The rise of elocution—continued', p. 5.

49 Campbell, Lawrence 1913, 'The elocutionist and the actor', *Theatre Magazine*, 1 November 1913, p. 6.

50 Maxwell, Stephen 1994, *The History of Soapbox Oratory: Part One: Prominent Speakers of the Sydney Domain*, self-published, Chiswick, p. 23.

51 Cited in Morrisson, Mark 1996, 'Performing the pure voice: elocution, verse recitation, and modernist poetry in prewar London', *Modernism/Modernity*, Vol. 3, No. 3, p. 32.

52 Besant, R. E. 1938, 'What is elocution?', *Age Literary Supplement*, 29 October 1938, p. 3.

53 Edmonds, Paul 1937, 'The art of speaking verse', *Australian Musical News*, 1 January 1937, p. 8.

54 Johnson, Lesley 1988, *The Unseen Voice: A Cultural Study of Early Australian Radio*, Routledge, London, p. 47; also see p. 71.

55 Peter Hay, cited in McKenry, 'Folk Poetry and Recitation', p. 151.

6. World English? How an Australian Invented 'Good American Speech'

Desley Deacon

In May 1922, Windsor P. Daggett wrote admiringly in the American entertainment magazine *Billboard* about the preaching of the Reverend Frederick W. Norwood, who was visiting from City Temple, London. 'He surprised the congregation,' Daggett observed, 'by talking "just like an American". At least his speech was so free from any trace of regional dialect that some of the audience was a little surprised to think that Mr Norwood was an Englishman.' 'Not a British intonation entered into his preaching,' Daggett went on, 'and his general impression of quiet force and normal speech was so similar to that of the regular pastor, Rev. Charles E. Jefferson, that one noted very little difference in their habits of speech.' Curious to know how Rev. Norwood had rid himself of his British 'accent', Daggett went to visit him, and concluded after talking to him that 'he never had any accent'. 'He comes from Australia,' Daggett discovered, 'where he has lived most of the time in the cities of Melbourne and Adelaide. The cultured speech of these cities appears to be the standard speech that has no local earmarks.' 'That is the test,' Daggett concluded, 'to be free from local peculiarities.'[1]

Commenting a few months later on a letter in the *New York Times* suggesting that the best speakers in England were now imitating the American accent, Daggett used the Rev. Norwood again as an example of good speech that avoided what he called the 'affected ultra-British, class dialect' of the 'la-de-da Britisher', which was no longer accepted as cultured speech and standard English. 'When this letter writer is told that the English are "imitating [the] American accent," he is uninformed,' Daggett concluded:

> The Englishman is simply dropping his class dialect and mannerisms of speech, and is adopting the standard of pronunciation that is widely accepted by the 'best speakers' in England and in the English-speaking world. When Rev. Frederick W. Norwood, of London, preaches at the Broadway Tabernacle, New York, and talks 'just like an American', he is not 'imitating' American speech. He talks as he has been accustomed to talk in Melbourne, Australia, and in London, England. He speaks English, standard English, with no local flourishes.[2]

Daggett was writing in his regular column in *Billboard*, 'The Spoken Word'. Week after week, from 1921 to 1926, Daggett commented in detail on the speech of actors appearing on the New York stage and of visiting lecturers and preachers such as Rev. Norwood. With all the zeal of the converted, he proselytised for what he called 'Standard English'. 'Standard English' was speech—like Norwood's—that was free of 'regional dialect'. 'There is a standard of English which can be heard in good society over the world so that it sounds as familiar and normal in Melbourne as it does in London or New York City,' he wrote. 'The actor in straight parts should be a master of that speech, and the preacher makes friends by speaking a universal language.' 'Considering that Mr Norwood fitted so naturally into Dr Jefferson's pulpit, I am bound to conclude that the London audience listening to Dr Jefferson at City Temple will say, "He doesn't sound a bit like an American," which is very much the sort of remark to be desired.'[3]

Daggett (1877–1958) came from an old New England family. After graduating from Brown University, he moved from instructor to Professor of Speech at the University of Maine. After serving briefly in World War I, he moved to New York, where he was, from 1927, Professor of Speech at the Hebrew Union College. He was also, for many years, theatre critic for *The Boston Transcript*.[4] In 1920, he enrolled for graduate study at Columbia University, where he came in contact with William Tilly, with whom he studied throughout the next decade.[5] Like many others who encountered Tilly, Daggett became a fervent disciple. His *Billboard* (and later *Theatre Monthly*) columns were an advertisement for Tilly's ideas of a universal English language.[6]

Tilly (1860–1935) was arguably the most influential speech instructor in the United States in the first half of the twentieth century. He and his students (or rather disciples) decided how Americans spoke to the world through theatre and through film. Tilly was not an American; nor was he British. He was, like the Rev. Norwood, an Australian who began his career as a teacher at Ironbark, New South Wales, the site, as all Australians know, of 'Banjo' Paterson's well-known poem *The Man from Ironbark* and now the village of Stuart Town.[7]

He was born in 1860, the son of Charles Tilley, a city missionary in Sydney, and his Irish wife Elizabeth, who came from the distressed Irish Protestant gentry family of Edgerton. Ian Clunies Ross, son of William Tilley's sister Hannah, describes his mother in his memoirs as proud of a supposed noble ancestry, fiercely status conscious and possessed of a keen, progressive drive—characteristics that seem to have been shared by her brother William.[8]

William Tilly attended the highly regarded Fort Street School and, encouraged by the Hon. John Fairfax, deacon of the Pitt Street Congregational Church (and founder of the *Sydney Morning Herald*), he continued his studies at the University of Sydney, from 1879–80, specialising in languages.[9] (The Rev. Charles Badham, a classics scholar from Oxford, was Dean of the Faculty of Arts and probably taught Tilly; and the eccentric Etienne Thibauld, a Doctor of Letters from the University of Paris, was employed as Modern Languages lecturer from 1882.

He, in turn, could well have been a student of the other great reforming linguist of the time, Paul Passy.) After his year at the University of Sydney, Tilly taught for the New South Wales Education Department in Ironbark, Wellington and Dubbo. In 1890, at the age of 30, he took his growing family to Germany. There he studied at the University of Marburg under the phonetician Wilhelm Vietor, the pioneer of what is now known as immersion language-teaching techniques, and he learned from Vietor his revolutionary 'direct method' of foreign language study.

Language study during this period had been revolutionised by Henry Sweet (1845–1912), a prominent member of the European movement of the mid-1880s to reform modern language teaching. His radical *Elementarbuch des gesprochenen Englisch* (*Primer of Spoken English*) (1885), written initially in German for German students of English, was extremely influential. Highly regarded in Europe and president of the International Phonetic Association from 1887 until his death, Sweet had no university post in his native England until he was appointed to the newly established readership in phonetics at Oxford in 1901 at the age of 56. In 1890, he published a translation of his *Primer of Spoken English* and a *Primer of Phonetics*; and his *Practical Study of Languages* (1899) discussed in detail the theories and practices of teaching and learning languages.[10]

William Tilly established the Institut Tilly, first in Marburg, then in Berlin, where students came from all over the world for intensive and rigorous language tuition. Among his students was Daniel Jones (1881–1967), who became the pre-eminent British phonetician of the early twentieth century.[11] Tilly's Australian students included Margaret Bailey (1879–1955)—later the gifted and inspiring headmistress of Ascham School in Sydney, where she taught French by the direct method[12] —E. G. Waterhouse (1881–1977)[13] and A. R. Chisholm (1888–1981), professors of German and English at the Universities of Sydney and Melbourne respectively, who also followed his method of teaching, and Edmund Herring (1892–1982), later Chief Justice of Victoria.

Tilly and his sons were interned at the outbreak of World War I in 1914 and were able to get to New York in 1917.[14] From that time until just before his death in 1935, he taught English and phonetics at summer sessions and extension courses at Columbia University.[15]

Teaching as he did in extension and summer courses, Tilly attracted many high school teachers and actors. He is always described as a strict 'Prussian' disciplinarian, but he must also have been a charismatic figure, because he inspired great love and admiration in his students, who spread his ideas through the New York school system and acting schools. Daggett describes him in a 1923 *Billboard* article as holding his mostly female audience at Hunter College spellbound as he lectured on, of all things, the letter 'r'—an obsession the two shared. As Daggett concluded in his article, 'The teachers swear by Tilly, and they are going to knock the "r" out of New "York" and several other places.'[16]

Part of Tilly's innovation as a proponent of the direct method was to teach speech patterns based on the spoken rather than the written word—and to use the relatively new International Phonetic Alphabet developed by Henry Sweet to accurately describe each sound.[17]

His greatest contribution, however, was the championing of what he called 'World English'—what his American disciples called 'Good American Speech' (as opposed to 'General American', which derived from the mid-west). World English was the sort of English the Australian Rev. Norwood spoke: speech that did not follow any regional dialect, including that of the British upper class. This was an invented accent that was considered appropriate to 'educated', 'cultivated' and 'cultured' English-speaking people all over the world. It did resemble, however, some New England speech patterns and resembled closely the non-regional variety of educated British English that Daniel Jones described in his 1918 *Outline of English Pronunciation* and called 'Received Pronunciation' in his 1926 edition—a book whose many editions remained authoritative guides throughout the first half of the twentieth century.

Tilly and his followers considered World English not only more intelligible to more people all over the world than any other dialect, but more beautiful. Indeed, his student Marguerite De Witt, who taught at Wellesley College, used her books *Euphon English in America* (1924) and *Our Oral Word, As Social and Economic Factor* (1928) to evangelise for what she called 'Euphonetics'.[18]

Tilly arrived in New York just at the right moment. Dr Frederick Martin (1881–1950) had recently been appointed Director of Speech Improvement for New York's public schools, and the training and licensing of teachers in this new area was urgent.[19] When Martin left in 1921, there was a long struggle for his position, which was won eventually in 1928 when Tilly's protégé Letitia Raubicheck, chair of the elocution department at the prestigious Julia Richman High School, replaced the unlicensed Agnes Birmingham.[20] (Birmingham had co-edited the textbook *First Lessons in Speech Improvement* in 1922, which favoured General American over World English.)[21] An influential figure on the faculties of Hunter College and New York University, Raubicheck further professionalised speech teaching; she instituted school oratorical contests and regular radio broadcasts for schools; and she kept what she called the speech problems of New Yorkers before the public with regular reports in the press.[22] A typical story in the *New York Times* in 1948 was headlined 'Speech aid is given to 30,000 in schools'; and, three years later, '56,640 students treated'.[23] Another Tilly student, Sophie Pray, who was president of the William Tilly Association—which was a sort of Tilly fan club—wrote the textbook *Graded Objectives for Teaching Good American Speech* for use in teacher training. 'Good speech signifies the possibility of readier spiritual integration with, and membership in, the cultured group in which most of us want to live as citizens,' she wrote. 'Ignorance may be condoned, lack of dexterity may be excused, but faulty speech and foreign accent are indelible signs of social inferiority.'[24]

Tilly's ambitions for a World English fitted well with visions of internationalism or transnationalism, which flourished in the wake of the devastation of the Great War. New York intellectuals Randolph Bourne and Elsie Clews Parsons proselytised for the idea throughout the war, claiming that the United States had the potential to become what Bourne called in an influential 1916 article 'the first transnational nation'.[25] Broadway theatre was already considered transnational in this way, attracting the best actors from all over the world. The actors whose voices Daggett praised most fully in his columns were those whose regional accents had been sheared away by performing all over the world. One actor he particularly praised was a Scot who had performed, not only in London and New York, but also in India, China, the East Indies and Australia.[26]

This vision of transnationalism was held also by movie executives such as Walter Wanger, executive producer at Paramount's New York studios from 1924 to 1931. Wanger saw films as 'foreign offices', with the potential to bring about world understanding and world peace much more effectively than diplomacy. When sound was introduced to films in 1927, Wanger and others saw this literally as an opportunity to speak to the world, and it was imperative financially and culturally to find a voice the world would listen to and understand.[27]

Tilly's World English, with its deliberate lack of 'localisms', fitted perfectly with this transnational ideal. The British diplomat and spy Bruce Lockhart, who studied under what he called Tilly's 'Spartan and Pitiless' regime before World War I, asserted in his memoirs that Tilly 'had done more for Anglo–German friendship than any man living'.[28] His American disciple De Witt emphasised the diplomatic purpose of their work in the dedication of her 1926 book, *Our Oral Word*, illustrated top and bottom by drawings of London and New York:

> To Page and Bryce
> Ambassadors
> of
> Head
> and
> Heart
> Throughout
> the
> English-Speaking World
> and thereby
> World Statesmen of Goodwill
> Goodwill United by the Many Seas.

Tilly's influence on the speech used in theatre and film was enormous. At the Windsor P. Daggett Studio—'Home of the Spoken Word'—on West Seventy-Fourth and Broadway, Daggett gave classes in voice production and pronunciation and produced his Spoken-Word records for mail-order students, to whom he also supplied the International Phonetic Alphabet and written lessons.[29] He also lectured on 'our American speech and voice' under the auspices of *Theatre Arts Monthly*, along with influential theatre personalities John Mason Brown and Kenneth MacGowan.[30] From 1923, he taught what he called 'The Spoken Word' at Richard Boleslavsky's American Laboratory Theatre and in 1926 he was on the faculty of the International Theatre Arts Institute with Brown and Frederick Kiesler.[31]

Also at the American Laboratory Theatre was Margaret Prendergast McLean, Tilly's star pupil.[32] She was at the same time Head of the Department of English Diction at the Leland Powers School of Theatre in Boston.[33] In the next decade, she taught at the Carnegie Institute of Technology and the Cornish School, Seattle. In the late 1930s, she followed her colleague at the Lab School Maria Ouspenskaya to Hollywood, where she taught at Ouspenskaya's Acting School.[34] It was she who wrote the textbook *Good American Speech*, first published in 1927, which set out the ideology and practice of teaching by the Tilly method.[35]

McLean's most successful pupil at the Powers School was Edith Warman, who joined her at the Lab School and began her own apprenticeship with Tilly at Columbia University. Warman became a speech instructor at Carnegie Tech's theatre training program, where she established her reputation as the most eminent theatre speech trainer in America and wrote her influential textbook *Speak with Distinction*. When she retired from Carnegie in 1968, she was hired by John Houseman as a founding member of the faculty of the new theatre program at the Juilliard School in New York, where she trained another generation of actors and teachers before her death in 1981.[36] Her former pupil Timothy Monich is considered the reigning voice coach today, and his students use a much revised version of *Speak with Distinction*, which he has edited.[37] To return the circle to Australia, Timothy Monich, according to a recent article in the *New Yorker*, is the speech coach of Australian film star Cate Blanchett.[38]

Film scholar Miriam Hansen argues that American mainstream cinema developed a 'global vernacular' (what she calls 'an international modernist idiom on a mass basis'), whose transnational appeal derived from diverse domestic traditions, discourses and interests, including those of the cosmopolitan Hollywood community. 'Hollywood did not just circulate images and sounds,' she argues, 'it produced and globalized a new sensorium; it constituted…new subjectivities and subjects.'[39] In other words, American films affected how people around the world looked, how they felt, how they talked and what senses they privileged; and they did this by putting on the screen a set of behaviours that came, as it were, from nowhere and everywhere.

I would suggest that the theatre was an important precursor of the new, international, sensory culture Hansen attributes to film. Theatre was, from the nineteenth century, a global industry whose touring companies penetrated deep into local communities. (Richard Waterhouse has shown this beautifully in his study of American companies touring nineteenth-century Australia with *Uncle Tom's Cabin*.)[40] Theatre was an important resource for the international women's movement from as early as the Woman's Christian Temperance Union's campaigns of the 1880s; and female actors were early models of modernity.

What is more, many early films were transferred directly from the stage, and personnel—actors, producers, directors and writers—moved backwards and forwards between the two mediums. From the mid-1920s to the early 1930s, many films were made in New York, and stage actors and producers mixed socially with those in film. In addition, Hollywood producers such as MGM's boy wonder Irving Thalberg came regularly to New York to decide which Broadway shows to buy, to see what was fashionable and what different audiences enjoyed. They would often film the performance as a guide to the movie. And, of course, they often persuaded the Broadway stars to move into the movies. If we are going to talk about a global vernacular, we have to take Broadway theatre seriously as one of its important sources.

As Tilly's story reminds us, Australia (and perhaps the other British colonies) played an interesting role in defining and popularising a new sort of universal speech that suited this developing global vernacular. I noted earlier that Daggett admired most the voices of those actors and preachers whose regional peculiarities had been sheared off through having to make themselves understood and socially placed in a large and diverse number of countries. Australia was fortunately placed to produce such actors, as it was the centre of a geographically and culturally diverse theatre circuit that extended from Britain, through South Africa and the East, and through the Pacific to the United States.[41] In addition, its immigrant, upwardly mobile population flocked enthusiastically to elocution lessons, where they learned to speak, not like the 'la-de-da' British upper class, but in the new invented accent of World English.

The actress Judith Anderson is a good example of this: the youngest daughter of a deserted mother who ran a school shop in Adelaide, she excelled at elocution as a child and went into the theatre, where her voice—throaty, refined but not 'la-de-da', and expressive—was her fortune on Broadway and in London.[42] Daggett admired the speech of many of the Australasian actors he heard on Broadway, and he often picked out Anderson's voice as a model of universal English speech.[43]

Miriam Hansen saw the new global vernacular disseminated by Hollywood as deriving from diverse domestic traditions, discourses and interests, including those of the cosmopolitan Hollywood community. Australians were part of that cosmopolitan community right from the start, and the New York theatre—where many of them flourished—was often the filter through which their cultural influence was refined and passed on to the world. I want to suggest in this chapter that one of their major cultural influences in the late 1920s and early 1930s was providing for this new world medium an intelligible cultured voice that was not 'la-de-da': a voice that was not recognisably from anywhere, a voice that was from everywhere and nowhere.

ENDNOTES

[1] Daggett, Windsor P. 1922, 'The Spoken Word', *Billboard*, 27 May 1922, p. 28. For Norwood, see 'Religious Services', *New York Times*, 6 May 1922, p. 10; and 'Peace in the Pacific. Speakers Emphasize Good-Will Between Nations in the East', *New York Times*, 12 May 1922, p. 4.

[2] Daggett, Windsor P. 1922, 'The Spoken Word', *Billboard*, 1 July 1922, p. 28.

[3] Daggett, Windsor P. 1922, 'The Spoken Word', *Billboard*, 27 May 1922, p. 28.

[4] 'Windsor Daggett Dies. Professor of Speech at the Jewish Institute of Religion', *New York Times*, 2 July 1958, p. 20.

[5] Windsor Pratt Daggett enrolled in the Graduate Faculties (now the Graduate School of Arts and Sciences) in 1911–12, 1920–26 and 1929–30, but did not graduate. Information courtesy of Jennifer Ulrich, Associate Archivist, Columbia University Archives and Columbiana Library.

[6] Daggett, Windsor P. 1925, 'Voice of the Magic Word', *Theatre Arts Monthly*, vol. 9, no. 6, June 1925, pp. 366–76; 'The Lineage of Speech', *Theatre Arts Monthly*, vol. 9, no. 9, September 1925, pp. 597–604; 'The Sounds of English', *Theatre Arts Monthly*, vol. 10, no. 1, January 1926, pp. 25–31.

[7] Stuart Town is a town of about 300 people 34km south-east of Wellington and 380km north-west of Sydney. It is a service centre to the surrounding area, where sheep and cattle farming and orchards are the major enterprises.

[8] Thomson, Philip, 'Tilly, William Henry', *Australian Dictionary of Biography 1891–1939*; Mitchell, Bruce 1981, 'Clunies Ross, William John (1850–1914)', *Australian Dictionary of Biography*, vol. 8, Melbourne University Press, pp. 33–4; Schedvin, C. B. 1993, 'Clunies Ross, Sir William Ian (1899–1959)', *Australian Dictionary of Biography*, vol. 13, Melbourne University Press, pp. 448–51; Clunies Ross, Ian 1961, *Memoirs and Papers, with Some Fragments of Autobiography*, Oxford University Press, Melbourne.

[9] John Fairfax (1804–77) was president of the Young Men's Christian Association, a member from 1871 of the Council of Education and of the Legislative Council from 1874. See Fairfax, J. O. 1972, 'Fairfax, John (1804–1877)', *Australian Dictionary of Biography*, vol. 4, Melbourne University Press, pp. 148–9.

[10] MacMahon, M. K. C. 2004, 'Sweet, Henry (1845–1912)', *Oxford Dictionary of National Biography*, Oxford University Press, available from http://www.oxforddnb.com/view/article/36385 (accessed 22 October 2004).

[11] Collins, Beverley 2004, 'Jones, Daniel (1881–1967)', *Oxford Dictionary of National Biography*, Oxford University Press, available from http://www.oxforddnb.com/view/article/38315 (accessed 22 October 2004).

[12] Lundie, Margaret 1979, 'Bailey, Margaret Ann Montgomery (1879–1955)', *Australian Dictionary of Biography*, vol. 7, Melbourne University Press, p. 138.

[13] O'Neil, W. M. 1990, 'Waterhouse, Eben Gowrie (1881–1977)', *Australian Dictionary of Biography*, vol. 12, Melbourne University Press, pp. 389–90.

[14] *British Australasian*, 31 December 1914.

[15] Jackson, Joseph MLA 1935, 'Obituary of William Tilly', *Sydney Morning Herald*, 9 October 1935, National Library of Australia MS William Tilly; Wright, Frederick 1935, 'Late William Tilly', Letter to the Editor, *Sydney Morning Herald*, 9 October 1935, National Library of Australia MS William Tilly; Thomson, 'Tilly, William Henry'; Jones, Daniel 1935, 'Obituary of Tilly', *Le Maitre Phonetique*, October 1935; Knight, Dudley 1997, 'Standard Speech: The Ongoing Debate', in Marian Hampton and Barbara

Acker (eds), *The Vocal Vision: Views on Voice by 24 Leading Teachers, Coaches and Directors*, Applause, New York, pp. 155–84.

[16] Daggett, Windsor P. 1923, *Billboard*, 5 May 1923, pp. 39, 43.

[17] Tilly's one article was 'The Problem of Pronunciation', in A. Drummond (ed.) 1925, *A Course of Study in Speech Training and Public Speaking for Secondary Schools*, New York; see Dickinson McDowell, Elizabeth 1926, 'The Speech Teachers Declare Themselves: Review of A. M. Drummond, *Speech Training and Public Speaking for Secondary Schools: being the report of a special committee of the National association of teachers of speech'*, *The English Journal*, vol. 15, no. 1, January, pp. 84–5.

[18] De Witt, M. E. 1924, *Euphon English in America*, E. P. Dutton, New York; De Witt, M. E. 1928, *Our Oral Word, As Social and Economic Factor, With a Comprehensive Group of Old World Euphonetigraphs*, E. P. Dutton, New York, J. M. Dent, London and Toronto, cited in Knight, 'Standard Speech', pp. 160–1, 169, 170.

[19] 'Dr F. Martin Dead. Speech Authority', *New York Times*, 11 April 1950, p. 32; 'Licenses Granted. For Various Positions In The Elementary Public Schools', *New York Times*, 27 July 1916, p. 17; 'No Foreign Accents For Good Americans', *New York Times*, 30 March 1919, p. 57; 'Alien Accents In English. Schools' Attack On Foreign Ways Of Speaking As An Aid To Americanization', *New York Times*, 14 March 1920, p. Xx1.

[20] 'Mayor's Niece Loses', *New York Times*, 1 December 1921, p. 16; 'Hylan School Board Re-Elects Officials Opposed By Mayor. Meleney, McAndrew, Snyder And Jones Unanimously Chosen For Six More Years. Members Disavow Politics Explaining Their Votes, They Criticise Gov. Miller And Denounce Newspapers. Mayor's Niece Withdraws But Candidacy of Rival to Be Director of Speech Improvement Is Rejected by Board', *New York Times*, 9 February 1922, p. 1; 'Miss McNally Gets A Secret Hearing. Superintendents Reconsider Nomination Rejected By Education Board. Hylan Relative Applying Action Taken Not Revealed, But Miss McNally Smiles As She Leaves Private Meeting', *New York Times*, 18 February 1922, p. 23; 'Miss McNally Nominated', *New York Times*, 21 February 1922, p. 5; 'Delay School Nomination', *New York Times*, 9 March 1922, p. 6; 'Mrs Martin J. Kennedy', *New York Times*, 15 June 1963, p. 23; 'Elizabeth McNally Director Of Speech Improvement Till Marriage 1923 To Democratic Re Congress For Upper East Side. Recommend Two For School Posts', *New York Times*, 24 September 1927, p. 8; 'School Nomination For Hylan's Cousin', *New York Times*, 25 September 1927, p. E5; 'Hylan Cousin To Lose Her Post In Schools', *New York Times*, 11 May 1928, p. 15; 'Three Are Nominated For School Posts', *New York Times*, 24 September 1928, p. 21; 'School Posts Filled', *New York Times*, 27 September 1928, p. 41; 'School Pay Sought By Hylan's Cousin', *New York Times*, 6 October 1928, p. 26; 'Board To Consider Birmingham Claim', *New York Times*, 7 October 1928, p. 31; 'Rules On Niece Of Hylan. Dr Graves Holds She Was Illegally Promoted To A School Post', *New York Times*, 18 January 1929, p. 13; 'High Court Bars Plea Of Agnes Birmingham', *New York Times*, 10 January 1933, p. 41.

[21] Krapp, George Philip and Anna I. Birmingham (eds) 1922, *First Lessons in Speech Improvement*, Charles Scribner's Sons, New York. See Knight, 'Standard Speech', p. 165.

[22] 'Hunter Faculty To Speak. Representatives To Attend Seven Conventions During Vacation', *New York Times*, 16 April 1930, p. 24; 'Prof. [Henrietta] Prentiss Honored. Association Of Teachers Of Speech Makes Her Life Member', *New York Times*, 4 June 1934, p. 10; 'City Teacher Tests In Speech Codified', *New York Times*, 26 October 1938, p. 25; '79 School Orators Chosen In Contest', *New York Times*, 23 March 1930, p. N5; 'Decries Bad Diction In Pulpit, On Stage', *New York Times*, 18 November 1931, p. 26. See, for example, 'Today On The Radio', *New York Times*, 10 June 1930, p. 38, and 24 June 1930, p. 29; Tompkins, Richard 1935, 'Speech Flaw Curb Sought In Schools', *New York Times*, 5 May 1935, p. N3; 'Choral Speaking Urged In Schools', *New York Times*, 9 May 1937, p. N7; 'Women Excel Men In Enunciation Bee', *New York Times*, 16 May 1937, p. 37; 'Speech Correction Gains In Schools', *New York Times*, 5 March 1939, p. D10; 'Careless Speech Fought In Schools', *New York Times*, 17 March 1939, p. 20; 'Defective Speech Remedied At Clinic', *New York Times*, 26 November 1939, p. 53; 'Plan Career Conference', *New York Times*, 14 April 1940, p. 16; 'Students And Public Aided By Speech Clinic At Queens', *New York Times*, 28 July 1940, p. 34; 'Speech Work Grows At Brooklyn College', *New York Times*, 13 October 1940, p. 62; 'LeHigh Engineers Studying Speech', *New York Times*, 22 December 1940, p. 40; Mackenzie, Catherine 1943, 'Good Habits Of Speech', *New York Times*, 4 April 1943, p. SM28; 'Speech Training For Teachers', *New York Times*, 12 January 1947, p. E9; 'Schools To Check On Speech Oddities', *New York Times*, 1 June 1948, p. 25; '13 Hunter Seniors Aid Speech Clinic', *New York Times*, 7 June 1948, p. 21; 'Will Teach Puerto Ricans', *New York Times*, 23 June 1948, p. 9; 'Radio Station Starts "Bad Speech" Contest For A Program To Teach Better English', *New York Times*, 2 September 1948, p. 25; Williams, James A. 1949, 'Course Improves Business Speech', *New York Times*, 17 April 1949, p. F8; 'Seeks Better Speech', *New York Times*, 17 January 1950, p. 31.

[23] *New York Times*, 4 October 1948, p. 25, and 28 September 1951, p. 41.

[24] Pray, Sophie 1934, *Graded Objectives for Teaching Good American Speech*, E. P. Dutton, New York, pp. 5–6, quoted in Knight, 'Standard Speech', p. 161.

[25] See Deacon, Desley 1997, *Elsie Clews Parsons: Inventing Modern Life*, University of Chicago Press, Chicago; Bourne, Randolph 1916, 'Trans-National America', *Atlantic Monthly*, 118, July 1916, pp. 86–97.

[26] Daggett, Windsor P. 1922, 'The Spoken Word', *Billboard*, 1 April 1922, p. 28.

[27] See Deacon, Desley 2006, '"Films as Foreign Offices": Transnationalism at Paramount in the Twenties and Early Thirties', in Ann Curthoys and Marilyn Lake (eds), *Connected Worlds: History in Transnational Perspective*, ANU E Press, Canberra, pp. 129–46.

[28] Lockhart, R.H. Bruce 1933, *Memoirs of a British Agent*, Putnam's Sons, New York and London; and Lockhart, R.H. Bruce 1934, *Retreat from Glory*, Putnam's Sons, New York and London.

[29] Advertisement, *Billboard*, 5 January and 15 March 1924.

[30] Advertisement, 'Theatre Arts Lectures', *Theatre Arts Monthly*, vol. 9, no. 6, June 1925, p. 493.

[31] Advertisement, *Theatre Arts Monthly*, vol. 10, no. 1, January 1926, p. 1; Advertisement, *Theatre Arts Monthly*, vol. 10, no. 10, October 1926.

[32] Advertisement, *Theatre Arts Monthly*, vol. 10, no. 1, January 1926, p. 1.

[33] Knight, 'Standard Speech', p. 174.

[34] 'Margaret McLean, Author On Speech', *New York Times*, 30 June 1961, p. 27.

[35] McLean, Margaret Prendergast 1935 [1928], *Good American Speech*, New and Revised Edition, E. P. Dutton and Company, New York.

[36] Knight, 'Standard Speech', p. 174; 'Service for Edith Skinner', *New York Times*, 15 September 1981, p. D31.

[37] Mansell, Lilene and Timothy Monich (eds) 1990, *A New Approach to Good American Speech*, Applause Theatre Books, New York.

[38] Lahr, John 2007, 'Disappearing Act: Cate Blanchett Branches Out', *New Yorker*, 12 February 2007, pp. 38–45, esp. 42–3.

[39] Hansen, Miriam 1999, 'The Mass Production of the Senses: Classical Cinema as Vernacular Modernism', *Modernism/Modernity*, vol. 6, no. 2, April 1999, pp. 59–77; also in Williams, Linda and Christine Gledhill (eds) 2000, *Reinventing Film Studies*, Edward Arnold, London.

[40] Waterhouse, Richard 1990, *From Minstrel Show to Vaudeville: The Australian Popular Stage, 1788–1914*, University of NSW Press, Sydney.

[41] See Deacon, Desley 2006, From Sydney to Hollywood: J. C. W. Stock Companies and the World Theatrical Market, Australian Modernities Conference, 'Vernacular Performers and Consumers', December 2006, University of Queensland; and Deacon, Desley (forthcoming), 'Becoming Cosmopolitan: Judith Anderson in Sydney, 1913 to 1918', in Desley Deacon (ed.) with Penny Russell and Angela Woollacott, *Biography Across Boundaries: Transnational Lives*.

[42] See Deacon, Desley 2005, An Adelaide Girl in New York: Judith Anderson's Broadway Success, 1924, Generations of Feminist Studies Conference and History Trust of South Australia, June 2005, Adelaide.

[43] See Daggett, Windsor P. 1924, 'The Spoken Word', *Billboard*, 12 January 1924, pp. 39, 42 on Irby Marshall ('She has a woman's voice that one seldom hears in America. It is rich in tone, low pitched and aristocratic without being in any way "genteel" or affected') and Leonard Willey; for Anderson, see *Billboard*, 10 May 1924, p. 40 ('Miss Anderson's voice has never been more flexible and flowingly expressive').

7. 'The Australian Has a Lazy Way of Talking': Australian character and accent, 1920s–1940s

Joy Damousi

> The Australian has a lazy way of talking through closed teeth. Much remains to be done.[1]

In his 1930 text, *Australia*, W. K. Hancock was perhaps more attuned to speech than other Australian historians either before or since his publication. In discussing the descendants of Australian convicts, he observed astutely that one clue for identifying them was through language. We 'may suspect', he observes, 'that there has come down to us, by subtle hidden channels, a vague unmeasured inheritance from those early days'.[2]

Hancock believed that it was absurd to try to replicate the English accent and 'attempt the impossible task of impressing upon scoffing pupils Oxford English thrice removed'. Teachers, he advised, would do 'better to develop the resources of this [the Australian] legitimate accent'.[3] He described the Australian accent as 'thin and narrow in its range of tone', but 'expressive and pleasant to the ear…The Australian intonation has in it something of heat-dazzle in "the land of lots o' time".' What was this Australian speech?

> It is smaller and simpler than the vocabulary of middle-class Englishmen, for Australia does not tolerate forms of thought and expression (such as irony) which are perplexing or offensive to the average man; and has also rejected, almost at a blow, the beautiful names of an intimate countryside—fields and meadows, woods, copse, spinney and thicket, dale, glen, vale and comb, brook, stream and rivulet, inn, and village. But in their place there is the Bush and a new vocabulary of the Bush—billabong, dingo, damper, bushwacker…Here, surely, is new wealth, expressive of distinctive and vigorous life, material for an individual literature.[4]

He observed how Australian words had their basis in the past; many came from Aborigines, some from the gold-rushes and others 'are originals coined off-hand out of experience and a matter-of-fact humour'.[5]

A study of the discussions about Australian speech from the 1920s to the early 1940s provides an examination of the ways in which a preoccupation with the Australian accent became a means of exploring Australianness. This period amplified the paradox of loyalty—at once to nation and to empire. The imagined community of Britishness defined the nation, but the significance of the distinctive Australian contribution expanded gradually to overshadow the narrative of empire. Even though there were subtle signs of this shift early, the peculiar 'double loyalty' endured during the inter-war years and early into the World War II period.

This chapter looks first at how popular discussion about the Australian accent and speech was promoted within a firmly embedded notion that Australia was a part of the British Empire. The tensions this created are considered in relation to the development of an acceptable, distinctive Australian sound. Another key theme consistent throughout these discussions is the understanding of speech as a form of moral instruction: whether it be within an individual, a national or international forum, speech and its sound were discussed in this way, pointing to correct behaviour and enunciation as a statement about moral worth in collective and individual terms.

Australian accent: 'the detestable snuffle'

Since the nineteenth century, educationists, politicians, social reformers and intellectuals have discussed the importance of speaking well and how a distinctive Australian accent does and should sound. Visitors to the colonies especially observed the distinctiveness of this speech. Louise Meredith, who lived in New South Wales from 1839 to 1844, documented the ways in which 'those born after the parents arrive in the colony have the detestable snuffle. This is an enigma which passes my sagacity to solve.'[6] Richard Twopenny commented in 1883 how the locally born gave scant attention to language and speech. In the colonial girl, he observed, '[l]anguages and other accomplishments are either neglected or slurred over'.[7] Mark Twain observed how when he visited Ballarat in 1897, the English spoken there was 'free from impurities…It is shorter than ordinary English—that is, it is more compressed. At first you have some difficulty in understanding it when it is spoken as rapidly as the orator whom I have quoted speaks it…I handed him a chair, he bowed and said: "Q".' This reduced form of English hardly had a sound; it was 'very soft and pleasant; it takes all the hardness and harshness out of our tongue and gives to it a delicate whispery and vanishing cadence which charms the ear like the faint rustling of the forest leaves'.[8]

Such a description, however, was lost on other local commentators, who expressed despair at the Australian accent. The most common judgement levelled against the accent was that of laziness. At the turn of the century, one writer noted in the *Bulletin* how laziness was identified as the chief cause of various deficiencies:

> The habit of talking with the mouth half open all the time is another
> manifestation of the national 'tired feeling'. Many of the more typical

bumpkins never shut their mouths. This is often a symptom of post-nasal adenoids and hypertrophy of the tonsils; *the* characteristic Australian disease. Such speakers produce pseudo *b*, *m*, *p*, with the lower lip and the upper teeth.[9]

The South Australian accent, in particular, was a combination 'polyhybrid of American, Irish brogue, cockney, county, and broken English'. One feature of this was 'tongue-laziness', and an anxiety to 'communicate as much as possible by means of the fewest and easiest sounds'. This laziness was manifest in the clipping of sentences and in the slurring of sounds:

> The method of producing sounds with the tongue…the false palate, the vibrating column of air leaving by both mouth and nose and slurring into a sound by means of an antesyllable, requires much less care, exertion, and expenditure of breath than that of making 'clean' sounds with the open mouth and proper use of the tongue and larynx.[10]

This accusation of laziness continued well into the twentieth century. A debate in the *Argus* during the late 1930s brought out the purists. 'Aint it 'ot' wrote in 1938 that since his/her son began to attend school, s/he had noticed that he 'developed an appalling manner of speech, and definitely an accent'. 'I learn with dismay that we have no professor of speech in our universities…Now that most schools have the wireless, could we not have cultured talks on this matter?'[11] Respondents endorsed these comments, with arguments for better educational services.[12] The professor of Education at the University of Melbourne, Professor G. S. Browne, argued that Australians needed to develop a speech with distinctive characteristics of their own: it 'should be the King's English, pleasantly spoken and probably with that peculiar resonance which was characteristic of Australian voices', although there was no need for 'Australian speech to include the great number of ugly sounds which it did now'.[13] J. Sutton Crow of the University Conservatorium argued that '[w]hat Australians generally seem to suffer from may be called "lip, tongue, and jaw laziness", leading to a lack of clear enunciation and a mumbling and slovenly mode of speech'.[14] Alfred Hart, a judge of the Melbourne Shakespeare Society, believed that mistakes in pronunciation were due to laziness and 'sloppy vocalisation'.[15] Slovenly speech in children was especially 'deplored',[16] while some argued that the 'lazy mouth and slovenly tongue were noticeable wherever English was spoken'.[17]

How could laziness be a part of modernity? Such laziness in speech suggested a lack of discipline, a poor standard in communication and a lack of moral standing. The ideal was a controlled and disciplined expression of the English language, which suggested a fashioning of the self that was more acceptable to polite, middle-class society. In another context, Dipesh Chakrabarty observes the way in which in Calcutta, 'informal, and unrigorous conversations' in public aroused suspicion and criticism. The social practice of '*ada*'—long, informal conversations—attracted accusations of idleness.[18] In Australia, it was a national question: through speech, it was hoped that a distinctive Australian identity would be preserved, but one that maintained connection to the English family of languages.

Whether it was English or Australian, some thought simply that the Australian accent was getting worse. James Green of the *Bulletin* argued that in Australia the public schools 'are not doing much in the way of teaching the rising generation to speak King's English, and our educated classes are themselves tainted'.[19] There was much discussion in the 1930s about the decline of 'Australian English' and the need to improve the standard of speech. There were certainly class and gender issues related to these questions. It was observed in 1933 by the *Sydney Morning Herald* that boys, 'especially at an early adolescence, appear to have a contempt for correctly modeled speech, which they regard as effeminate'. Even those who did speak well were under suspicion. The 'clear-cut, well-modulated speech of a speaker of polished diction is regarded with suspicion, and even with hostility, by the majority, for the insufficient reason that he has the temerity to be different from the herd'.[20] The best way to bridge such class divisions was to develop a national pride in speech. In fact, the 'best speech' was that which obliterated class differences, but it seemed to keep gender differences alive:

> It is the sort of speech which will carry a man anywhere, which does not attract undue attention to itself, which is understood by the greatest number of people in every part of the country which suggests that the speaker is a man of decent education who is used to mixing with a variety of his kind and does not announce his birthplace or consign him to any class.

It was clear that the Englishman's definition was inapplicable, and an inadequate prescription for Australia.

> We would define Australian-English as that pleasant oral communication which is audible and instantly apprehended by reason of its clear enunciation and rate of articulation; which is expressed in correct grammatical form and is free from solecisms; it has the vowel quality and absence of nasality associated with a person of respectable attainments, and the inflections are such as do not provoke a sense of antagonism or resentment in the auditor by virtue of such speech.

Why did Australians slur their speech and have such problems with it? Climatic and environmental changes were identified as important, but above all, the reason for 'mumbling speech' was 'diffidence and even laziness'. The fight for 'good speech is a noble ideal, worthy of all ranks; the language is one that should stimulate the pride of our young and virile nation'.[21]

The debate about the Australian accent preoccupied educationists. In a lengthy engagement with the issues of the Australian 'accent', John K. Ewers argued in 1937 that although the Australian accent was 'a superior speech' to the purer English, we should be careful before deluding ourselves 'into believing that all is well with Australian speech'. Australian speech should 'not be left to develop in a haphazard way into something which is careless and slipshod'. Two methods were proposed: to get rid of the accent altogether, 'before it becomes too deeply rooted', or to maintain it as it was. Drawing on the conventional explanation that speech was determined environmentally by Australian conditions, Ewers argued that it would be impossible to improve the national Australian accent. Within these 'climatic and temperamental limitations', however, Australian speech should be enhanced.[22]

In other quarters, the teaching of elocution was identified as one means of enhancement. In his forthright publication *Correct English, Public Speaking, Elocution, Voice Production*, C. N. Baeyertz argued for a celebration of Australian accents; he recognised that Australians did not need to sound British, but nevertheless, they did not speak well. Indeed, for Baeyertz, this was a major tragedy.

> We are in the presence of tragedy; and tragedy is never less than quite serious…The tragedy of all this is that we Oss-stryke-yuns or Ozzies (as we are also apt to style ourselves) are usually in a state of ignorance (neither blessed nor in any sense blissful) as regards our general attitude to simple English speech, and that the depth of our ignorance is nowhere more murkily shown than in our habitual flattening of the easy and inoffensive vowels.[23]

There was a deficiency in listening and in sound. Most young children suffered from a defect of sound, he observed: 'Their faulty pronunciation and indistinct articulation are almost entirely due to inadequate or perverted development of sound-perception. Ears have they, but they hear not.'[24] It was in this auditory capacity that there was a deficiency. The inability to listen was a dramatic shortcoming:

> The ear of the average Australian child is obviously untrained. Bad teaching has so dulled his auditory sense that he no longer hears with distinctness and accuracy. It seems that he is never taught to listen to his own voice and detect his own errors of lips and tongue. In Australia, as a rule, the lips are little used in speech; they are immobile, mere obstacles in the way of sound-emission. They remain rigid, and behind them the Australian tongue is literally an unruly member.[25]

The accent itself had deteriorated: 'It is beyond question that the Australian accent is becoming vastly more aggressive every year. In nine cases out of ten, we do not speak as accurately as our fathers and mothers, and our children know little of, and often care less for, "the well of English undefiled".'[26]

The work and writings of the elocution teacher E. Stanley Brookes reflect the move during the inter-war years for an appreciation of the development of an Australian—and not simply the replication of a British—accent. Brookes was a defender of the Australian accent and he argued that it was the English and not Australian accent that was offensive. 'The vast majority of the early settlers came from England and brought their cockney and other accents with them. Therefore our unmusical and ugly Australian accent is more English than Australian.'[27] An elocutionist 'of marked ability', Brookes was an advocate of a standardised pronunciation in Australia, but it was important that an Australian accent be 'attractive, not repulsive'. What was the best speech? He believed it was that which was clear and simple, nothing too affected: the 'glorious heritage of our English language is simple speech'. It was 'right and natural' that 'we should have an Australian accent'.

An appreciation of the subtleties of the English language was one of the key issues that concerned international as well as local authors—as well as the need to maintain some purity of speech. Moral instruction was never far removed from the commentary. Alison Hasluck announced that this would not only develop self-control and power, 'but that they are necessary adjuncts to the fullest expression of the subtleties and beauties of the English language'.[28] The sound of language and speech also posed a problem, as she saw it, and there was an urgent need to train the ear. The ear and organs must be trained by viva voce examples and by practice. Imperfect or wrong production of sound was often associated with tricks that marred facial expression as well as pronunciation. 'Immobility of the lips, a twist or contortion of the lips, a twist of the jaw, clenching the teeth, moving one side of the mouth more than the other—these faults should be dealt with.'[29]

The bodily and social benefits of elocution were also considered important, so that correct speech became a branch of deportment and etiquette. Kathleen Rich, in her *The Art of Speech: A Handbook of Elocution*, published in London in 1932, identified the benefits of elocution in terms of 'gaining self-confidence and poise in speaking, not only in public and on the stage, but in social intercourse'; it was 'undoubtedly a benefit to health, especially to the nervous system, and to the chest and throat'.[30] Victor MacLure, in his 1928 elocution manual, which was distributed internationally throughout the Empire—in London, Bombay and Sydney—identified the need for elocution because of poor speech and the offensive use of it socially. He observes:

> The shrill voice, no matter how clever or amusing the ideas it expresses, must in the end become an irritation. The indistinct speech, no matter how sound or revealing the thoughts it embodies, most surely will be found exhausting in the long run. One's ears cannot be absorbing ugly sound or straining to catch mumbled phrases for any length of time without some physical reaction. Think, indeed, how much the pleasant, musical, well-modulated voice is an asset in personal appeal.[31]

Learning to speak was therefore most valuable: 'By practice of reading aloud, of speaking up and out, of placing your vocalised breath rightly in your mouth, your voice takes on a brighter ring. It becomes pleasanter to hear.'[32] There was a danger of losing the strength, vitality and originality of how English should be spoken unless this decline was corrected. Here again, the concern had cultural implications: 'It is time that something was done to arrest this decay in English speech...Your French is a snob. Your Spanish is a hidebound hidalgo. Your German is something of an unreceptive boor. But your English is a gentleman of easy manners who finds good in everything.'

The threat, however, was not from without; it was in fact from within a slothful and slovenly attitude to speech.

> There is no danger of English losing its character through infusion of alien words. Time and again it has been completely invaded, only to emerge stronger than ever. If there is a danger threatening our language it is that it may lose its character through slovenly use in speech. We are become [sic] a nation of mumblers. Lispers, gutturalists, ventriloquists, mouth-breathers, butchers of sweet sounds. And our carelessness is threatening our most valuable heritage.[33]

Like MacLure's book, Alexander Watson's *Speak Out! The Commonsense of Elocution* (1924) was published in London and circulated in Calcutta and Sydney. Elocution became part of the endeavour to speak well; it was pitched primarily as training for public speaking. The emphasis on the strained gesture and bodily movements are gone, but the stress on correct enunciation in conversation, distinct utterance and so on frame the instruction. Watson's aim in the volume was the 'clearness of voice and verbal audibility...[to] render it useful to a very wide public, and stimulate many to take pride in well-spoken English'.[34] The book was based on Watson's lectures to students when he was lecturer on the speaking voice at Birkbeck College, a post he relinquished to 'fulfil a long series of engagements in Australia, New Zealand, and the United States'. The lectures were also redelivered at Westminster College in Cambridge, among other places.[35] For Watson, it was not appropriate for correctness to be 'pedantic, formal, or pernickety'. This was probably as bad as, if not worse than, bad speech:

'Anything like an exhibition of obvious virtuosity, particularly in conversational speech, would be abhorrent.'[36]

The teaching of a particular type of English dominated Australian schools. Jill Ker Conway recalls how the education she received in the 1930s and 1940s stressed this emphasis on speech, deportment and etiquette. When she enrolled in the exclusive Sydney girls' school Abbotsleigh, she found there was emphasis on English literature especially, and on speech and enunciation.

> Our curriculum was inherited from Great Britain, and consequently it was utterly untouched by progressive notions in education. We took English grammar, complete with parsing and analysis, we were drilled in spelling and punctuation, we read English poetry and were tested in scansion, we read English fiction, novels, and short stories and analysed the style. Each year, we studied a Shakespeare play, committing much of it to memory, and performing scenes from it on April 23 in honor of Shakespeare's birthday...This gave us the impression that great poetry and fiction were written by and about people and places far distant from Australia...to us poetry was more like incantation than related to the rhythms of our own speech.[37]

Speech and deportment were central aspects of this education. 'Speaking loudly, sitting in public in any fashion except bolt upright with a ramrod-straight back, were likewise sorts of behaviour which let down the school.'[38] Speech became part of deportment:

> One's voice must be well modulated and purged of all ubiquitous Australian diphthongs. Teachers were tireless in pointing them out and stopping the class until the offender got the word right. Drills of 'how now brown cow' might have us all scarlet in the face choked [with] schoolgirl laughter, but they were serious matters for our instructors, ever on guard against the dipthongs that heralded cultural decline.[39]

One of the aspects of speech was that it was regarded as a continuous source of moral instruction, for women in particular. The cultivation of voice signifies a transition from boyish rowdiness to mature women. The advice in women's journals points to the importance of voice, speech and culture in the development of femininity. Over several years, various advice columns were run in *Everylady's Journal*. The following advice was offered to 'Girls of the Sunny South' by Domina:

> Maturity pardons the rough speech used in the rough games, the loud voice, the boyish, ugly stride, the general tone and bearing—pardons, condones these things in the youthful...The woman of maturity and charm...hears with critical ears the crude, direct language. And hearing, condemns; she listens to the loud mannish voice, and more carefully

modulates her own. She notes the masculine bearing, and resolves immediately that the price paid for prowess in rough games such as hockey, etc., is far in excess of values received.[40]

It is important in youth to begin to correct and modify early boyish behaviour: 'Understand that the woman who speaks with an intonation delightful to hear, has, in her youth, watched her utterance, and carefully guarded it from crudity and a directness which the world…regards as deplorable!'[41] The importance of correct speech was the subject of an address given by Helen Munro-Ferguson, the wife of the then Governor-General, to a girls' college in NSW. Inspiration and trust can be drawn

> if you respect the language and endeavour to speak it with purity of diction and accent, avoiding stupid, imported slang and the habit of making one word, such as 'awful', do the work of half a dozen. It has been said that the quality of a man's brain can be gauged by the adjectives he uses, so remember that, if you happen to state that this or that is 'rotten', when you merely mean that it is 'tiresome' or that a person is 'decent' when you mean she is 'kind', an attentive hearer will take a depressing view of your mental powers, and perhaps be reminded of the fairy story and the beautiful Princess, from whose mouth a toad hopped every time she spoke.[42]

Throughout the 1920s and 1930s, such advice escalated and was a regular feature of the journal on conversation, voice and presentation of women's sounds. It was not only women, however, who were given advice. In 1925, advice was given to a male reader who needed to be better versed in the art of 'correct speaking':

> I think you would find a course of elocution of great assistance to you. Careless pronunciation, which seems to be on the increase nowadays, would seem to be at the root of your difficulty. If an opportunity occurs to mentally pronounce a word of many syllables before uttering it aloud, by all means take advantage of it. Each syllable has its own value, and is placed in the word for a special purpose—to be pronounced distinctly.[43]

Polite talking and conversation were identified as other aspects that were key to femininity. Good conversationalists

> are born not made…To be an entertaining conversationalist, it is necessary to have a good, all-round knowledge of present-day affairs, to be conversant with the latest play or book and to know just sufficient about it to make an intelligent reference…It is little use memorizing a set conversation, as it might never be used; moreover, it is always well to adapt yourself to circumstances…Above all, be natural. People who

adopt an affected, mincing manner of speech look perfectly ridiculous when caught off their guard, and it is better by far to play the part of an intelligent listener than to indulge in a stream of wearisome small talk.[44]

'There is nothing wrong with Australian speech'

These debates were heightened even further in 1942, when the question of Australian speech aroused the interest of ABC listeners and was debated in the *ABC Weekly*. After the advent of radio, the debate about the correct way of speaking on the radio was a source of much discussion. As Ken Inglis has shown, the BBC radio voice was favoured on Australian radio for much of the 1930s and 1940s.[45]

In 1942, A. G. Mitchell, an academic in the Department of English at the University of Sydney, who would later be first Vice-Chancellor of Macquarie University, wrote two articles in the *ABC Weekly* that discussed the merits of Australian speech. He was strident in his view that an Australian voice should be broadcast on the ABC: 'We should use an Australian speech, without apology and without any sense of a need for self-justification. There is *nothing* wrong with the Australian voice or speech. It is as acceptable, as pleasant, as *good English* as any speech to be heard anywhere in the English-speaking commonwealth.'[46]

He opposed Australians trying to imitate the English manner, which would result in producing 'a speech that was neither good Australian nor good English…We should give the pronunciation that is commonest in Australia, not slavishly imitate the English pronunciation'.[47] Judgements are never too far away from these discussions. 'Few prejudices,' observes Mitchell, 'are more easily aroused than those that concern variations in speech.' The conflation between speech and morality is drawn out in Mitchell's article:

> Question a man's pronunciation of a word and you may touch him as nearly as if you doubted his moral integrity. Differences in political opinion are often more readily tolerated than differences in pronunciation. We are prepared to believe that a man who differs from us in politics may still be a quite reasonable person. But many of us go through life in the comfortable faith that any man who speaks differently from the way in which we speak must be a knave or a fop or a chump.[48]

Mitchell argued for tolerance when confronted with differences of language and accent. 'The tolerance we should feel towards people of other English-speaking countries we might reasonably claim for ourselves. Visiting critics, whatever their qualifications or interests, have a constant habit of condemning Australian speech, often by invitation from interviewers.'[49] To 'read their opinions one would think that Australians had a monopoly of everything deplorable, careless and corrupt in pronunciation'.[50] In defending Australian speech, Mitchell was questioning the practice of the ABC at that time and presenting a challenge to the policy of the national broadcaster, which was to promote a middle-class, educated English voice.

The *ABC Weekly* editorialised that it 'embraces the problem of whether broadcasting is to exert its proper influence over listeners by sounding natural and attractive to them or whether it is to be discounted by them as aloof and superior'.[51] The magazine was definite that the voice that was broadcast should, however, be Australian. One of the 'depressing' tendencies about broadcasting was that there was a tendency to 'standardise' speech. 'We have suffered from the imitation of the so-called BBC voice in this country, just as the BBC has suffered from imitation in its own ranks.' It was important to resist the 'suburban fear' that somehow the concern about being considered '"uncultured" shall trick us into imitation of something which can never be anything but imitation and therefore of little value'.[52] On another occasion, it was argued that the selection of broadcasters should be made on the basis of how well they spoke 'the Australian brand of English well', but 'not those who merely imitate something foreign to our environment'. They would then make the speech 'more pleasant to the ear'. A separation from the speech of 'ordinary' people 'merely alienates the people and sets up a natural resentment'.[53]

Not all listeners agreed. Some argued that there were many Australian voices and accents. 'Another Australian' believed that 'most Australian voices have a nasal intonation which does not come through the microphone at all well, some words being quite unintelligible'. While it was important to have Australian broadcasters, they needed to be taught 'to speak the King's English'.[54] The Australian voice obviously had a number of variations. 'After listening to the Australian voices heard round me all day,' wrote one correspondent, 'I find the delivery and accents of the ABC announcers heard at night quite refreshing and pleasing to the ear.'[55]

These articles generated considerable debate and discussion. Some argued that there was no such thing as an Australian accent, while others argued that if Australians insisted on speaking like they did, then 'we must and will do so, but we can hardly expect respect for a language bred of carelessness out of ignorance—still less to find it considered "as acceptable, as pleasant, as good English"'.[56] The slovenly nature of Australian speech remained a source of considerable concern throughout the 1940s. 'As a mother of five young hopefuls,' wrote 'Speech' from Toowoomba, 'I find there is a tendency to slur the vowels—a fault which shows up when they are asked to spell the word in question. This makes for slovenly speech and needs watching.'[57] Others were appalled at the suggestion that Australian speech was preferable to what was understood as standard English: 'National arrogance and conceit can go no further than to claim that an untrained Australian voice is superior to that which results from study and hard work...Please do not degrade the cultural level of ABC announcers.'[58]

Some saw it as a choice between buying 'a shoddy, second-class material product if it was within his financial possibility to buy a first quality article'. Why should the cultivation 'of the best mental or educational production of a language...be despised?'.[59]

Prominent Australian commentators were asked to contribute to a debate about whether it was desirable to imitate the English or to develop and encourage a distinctly Australian accent. R. G. Menzies, a royalist and devotee of the British, did not object to an Australian speech, with 'local colour', but he detested the 'widespread slovenliness of speech'. 'I am all against encouraging carelessness and indifference on these matters. We will be none the less good Australians by being a little fastidious in our expression and using words as if we really knew what they meant.'[60] Vance Palmer, a radical nationalist, was more strident in his insistence that there be an Australian voice adopted and celebrated. It is beyond argument, he maintained,

> that the standard we ought to aim at is good Australian speech, not any other kind. It may not have the richness of good Irish, the rhythm of good American, or the tonal variety of good southern English; but it has its own quality. Anyhow, it is native to us: that ought to clinch the matter. Any effort to substitute something else would lead to the enthronement...ringing emptily like a vessel with nothing inside.[61]

In an article entitled 'There is No Australian Accent', Dal Stivens identified various characteristics as quintessentially Australian in our speech. Citing a Sydney teacher of voice production, Stivens argues that much of the 'slovenliness in Australian speech is a psychological inhibition'. He argued that: 'The Australian hates to be conspicuous or "different". He has a horror of "side". Every teacher finds himself opposed, sooner or later, by what has been termed "a sturdy reluctance to vary from what is considered 'natural speech'".'

Historically, Australia had also been a ruggedly masculine country: conquering the bush meant there was 'a distrust of ideas and the things of the mind', although this was changing gradually. With urbanisation and increased communication, this has changed. Quoting the voice teacher, Stivens notes the way in which this teacher argues that the

> so-called Australian accent is the product of a highly strung and nervous race. The Anglo-Saxon is the most repressed of all races. The Australian...[bears] the shame of his convict beginnings: he is subservient to England and to English customs. He had what the popular psychology calls an 'inferiority complex'.

This state was apparently reflected in speech: all pupils spoke with tight lips, did not open their mouths, were diffident and inclined to bluster.[62]

Discussion about the sound of accents invariably led to considerations of the Australian national character. Reflections on what this constituted drew out various stereotypes and clichés, but the label of laziness and slovenly speech consistently informed such discussions. Debates about received English compared with the local variant and questions relating to the cultural politics of moral self-improvement are central themes in these debates. The most significant shift during the inter-war years was, however, the perception and promotion of a cultivated Australian accent as opposed to one that was considered rough and slovenly. By this time, it was agreed that the English accent should not simply be copied or replicated; the challenge ahead was how a respectable and distinctive Australian sound should be spoken.

ENDNOTES

[1] *Argus*, 11 December 1935, p. 21.

[2] Hancock, W. K. 1930, *Australia*, Jacaranda, London, p. 28.

[3] Ibid., p. 252.

[4] Ibid., p. 252.

[5] Ibid., p. 252.

[6] Meredith, Mrs Charles 1973 [1844], *Notes and Sketches of New South Wales*, Penguin, Melbourne, p. 50.

[7] Twopenny, Richard 1973 [1883], *Town life in Australia*, Penguin, Melbourne, p. 84.

[8] Twain, Mark 1973 [1897], *Mark Twain in Australia and New Zealand*, Penguin, Melbourne, pp. 237–8.

[9] 'The Red Page', *Bulletin*, 23 March 1901.

[10] Ibid.

[11] *Argus*, 1 December 1938, p. 7.

[12] *Argus*, 3 December 1938; 6 December 1938, p. 33.

[13] *Argus*, 3 November 1936, p. 8.

[14] *Argus*, 9 November 1936, p. 8.

[15] *Argus*, 29 October 1936, p. 10.

[16] *Argus*, 21 December 1936, p. 3.

[17] *Argus*, 12 May 1937, p. 11.

[18] Chakrabarty, Dipesh 2000, *Provincializing Europe: Post-Colonial Thought and Historical Difference*, Princeton University Press, Princeton, pp. 180–213.

[19] Green, James 1926, 'The Australian Accent', *Bulletin*, 14 October 1926, pp. 3–4.

[20] *Sydney Morning Herald*, 28 December 1933, p. 6.

[21] Ibid.

[22] *Argus*, 10 December 1937, p. 10.

[23] Baeyertz, C. N. c.1935, *Correct English, Public Speaking, Elocution, Voice Production*, C. N. Baeyertz Institute, Sydney, NSW.

[24] Ibid., p. 6.

[25] Ibid., p. 7.

[26] Ibid., p. 75.

[27] *Argus*, 30 October 1936, p. 10.

[28] Hasluck, Alice 1925, *Eloctuion and Gesture*, Methuen & Co., London, p. viii.

[29] Ibid., p. 47.

[30] Rich, Kathleen 1932, *The Art of Speech: A Handbook of Elocution*, London, p. xii.

[31] MacLure, Victor 1928, *The Practical Elocution Book*, London, Bombay, Sydney, p. 19.

[32] Ibid., p. 19.

[33] Ibid., p. 20.

[34] Watson, Alexander 1924, *Speak Out! The Commonsense of Elocution*, London, p. 8.

[35] Ibid., p. 8.

[36] Ibid., p. 42.

[37] Conway, Jill Ker 1989, *The Road From Coorain*, Knopf, New York, p. 98.

[38] Ibid., p. 101.

[39] Ibid., p. 102.

[40] *Everylady's Journal*, 6 April 1918, p. 230.

[41] Ibid.

[42] *Everylady's Journal*, 6 October 1919, p. 565.

[43] *Everylady's Journal*, 7 December 1925, p. 1050.

[44] *Everylady's Journal*, 1 August 1929, p. 138.

[45] Inglis, Ken 1983, *This is the ABC: The Australian Broadcasting Commission, 1932–1983*, Melbourne University Press, Melbourne, p. 70.

[46] *ABC Weekly*, vol. 4, no. 37, 12 September 1942, p. 3.

[47] Ibid., p. 4.

[48] *ABC Weekly*, vol. 4, no. 36, 5 September 1942, p. 3.

[49] Ibid.

[50] Ibid.

[51] *ABC Weekly*, vol. 4, no. 37, 12 September 1942, p. 10.

[52] *ABC Weekly*, vol. 4, no. 29, 18 July 1942, p. 10.

[53] *ABC Weekly*, 1 August 1942, p. 10.

[54] *ABC Weekly*, vol. 4, no. 30, 25 July 1942, p. 4.

[55] Ibid.

[56] Ibid., p. 4.

[57] *ABC Weekly*, vol. 4, no. 4, 10 October 1942, p. 2.

[58] *ABC Weekly*, vol. 4, no. 35, 29 August 1942, p. 2.

[59] *ABC Weekly*, vol. 4, no. 34, 22 August 1942, p. 2.

[60] *ABC Weekly*, vol. 4, no. 39, 26 September 1942, p. 11.

[61] Ibid.

[62] Stivens, Dal 1940, 'There is No Australian Accent', *Home*, vol. 21, no. 3, March 1940, p. 43.

8. Towards a History of the Australian Accent

Bruce Moore

In 1927 in *Australian Pronunciation: A Handbook for the Teaching of English in Australia*, Ruby W. Board presents a firm notion of what the best pronunciation of English is:

> In every English-speaking country there is to be found amongst cultivated people a certain pronunciation, which is unconsciously accepted as the best speech. On examination no trace of dialect can be detected, nothing that will single out the speaker, no touch of provincialism or of affectation. It is understood by all without effort, it is pleasant to the ear, and it may be heard in England, Scotland, Ireland, in all of the British Dominions, and even occasionally in America. It is significant that while the particularly obtrusive quality of the Oxford man, the Cockney, the Yorkshireman, the Scotchman, the Australian, the Canadian, is noted and labelled, the speech referred to above is described as *good English*.[1]

The swipe at Americans and American speech demonstrates the extent to which this cultivated pronunciation (or 'received pronunciation', as it had been called by A. J. Ellis in 1869)[2] is a product of empire. Its endorsement as 'the best speech' at this point in Australia's history is entirely in keeping with the dying out of the radical nationalist movement in the 1890s, and the development of the curious blend of Australianness and Britishness that framed notions of nationalism in twentieth-century Australia. As Ben Wellings argues: 'Much of the content of Australian nationalism was derived from Britishness and adapted to local conditions, resulting in national consciousness as much merged with imperial symbols, as developed from the vernacular culture of the Australian people.'[3] Another educationalist, Ethel M. Mallarky, writing in 1914, attempts to have it both ways when, on the one hand, she accepts the fact that Australian English has diverged from 'accepted standards of speech' ('It is to be expected—and need we strive to hinder it?—that a distinctive Australian speech will emerge with the Australian people's emergence into nationhood'), but goes on to insist that standards of empire pronunciation must be maintained:

> But there are certain grave dangers by which the English tongue is being assaulted in Australia, peculiarities which may seriously militate against its efficiency as an instrument for the communication of thought. And further, as habit of speech tends to mark social boundaries, so in a wider

degree, and far more distinctly, may it operate to demarcate national boundaries and lead to a weakening of that common sentiment which maintains healthy life in the Empire, and serves to keep in sympathetic relations all the widely scattered units of the English-speaking race.[4]

In this environment, it is not surprising that about this time the Australian accent became a target for quite widespread criticism. Valerie Desmond in *The Awful Australian* (1911) writes:

> In conclusion, it is only necessary to point out that so objectionable is the Australian accent that theatrical managers resolutely refuse to employ Australian-born actors and actresses. Though a few of these are possessed of talent—or what passes for talent in Australia—the managers prefer to import English artists of inferior merit, solely because they possess the essential qualifications that Australians lack—the ability to speak the English language.[5]

In the same year, William Churchill writes: 'The fact remains that the common speech of the Commonwealth of Australia represents the most brutal maltreatment which has ever been inflicted upon the language that is the mother tongue of the great English nations.'[6]

In 1926, a former Director of Education in New South Wales bemoans the accent: 'If we must follow a dialect of English in Australia, why not follow one of the charming ones? Why follow the ugliest that exists?'[7]

In the nineteenth century, there were some occasional early comments on the Australian accent that were critical, or that might be taken as critical. First, there is Peter Cunningham, who, in *Two Years in New South Wales* (1827), claims that 'the London mode of *pronunciation* has been duly ingrafted on the colloquial dialect of our Currency youths, and even the better sort of them are apt to meet your observation of "A fine day", with their *improving* response of "*Wery* fine indeed!"'.[8] There is, however, no other supporting evidence for Cunningham's claim of Cockney influence, and it is not until the end of the century that the Cockney shibboleth appears again—at a time when there is no longer a possibility of influence.

Secondly, there are writers who comment on intonation and articulation. Mrs Charles Meredith, in *Notes and Sketches of New South Wales During a Residence in the Colony from 1839 to 1844* (1844), claims that 'a very large proportion of both male and female natives *snuffle* dreadfully; just the same nasal twang as many Americans have'.[9] Nasality is commented on by R. H. Horne in *Australian Facts and Prospects* (1859),[10] and by a number of writers in the 1890s. Rolf Boldrewood in *Robbery Under Arms* (1882–83) says that in contrast with the Americans, 'most of the natives have a sort of slow, sleepy way of talking'.[11] The Australian-born Rosa Praed also notes the elision of consonants in *The*

Romance of a Station: An Australian Story (1889) in the speech of the stockman Tillidge: 'He pulled up, nodding to Alec's "Good-day, Tillidge," and replying in a short, morose manner, running his words one into the other, as a bushman does, "G'd-day, sir"'.[12] In *Outlaw and Lawmaker* (1893), Praed notes a drawl in Australian speech: 'Lady Horace came slowly down the log steps, and held out her hand to Hallett. "How do you do," she said, in her gentle little Australian drawl.'[13]

Interestingly, these features—nasality, a flatness arising from a lack of variation in intonation, a drawl, elision of syllables—became the stereotypes of Australian English in the first half of the twentieth century,[14] but from the evidence we have they were hardly of great concern in the nineteenth century. Indeed, the typical descriptions of the Australian accent for most of the nineteenth century are not critical, and do not focus on what might be potentially negative aspects of the accent; they are overwhelmingly positive. James Dixon, in *Narrative of a Voyage to New South Wales and Van Dieman's Land in the Ship 'Skelton' during the year 1820* (1822), writes: 'The children born in those colonies, and now grown up, speak a better language, purer, more harmonious than is generally the case in most parts of England. The amalgamation of such various dialects assembled together, seems to improve the mode of articulating the words.'[15]

Especially in the light of recent research on the development of colonial English, this is the most significant observation on Australian English in the nineteenth century. New Zealand researchers[16] have shown—from an examination of recordings of New Zealand speakers who were born to the first generation of settlers in New Zealand—that the children of this first generation did not reproduce the accents of their parents, and did not develop a single accent under peer-group pressure. Rather, each child picked up different features of accent from the varying dialects surrounding them. It is in the following generation that the 'foundation accent' is established, when all of the children speak with much the same accent. Due to the special circumstances of NSW being a penal colony, it is not possible simply to transport the New Zealand pattern to Australia, but whether Dixon was listening to first-generation or second-generation children, the significant fact is that he notes that their speech is not heavily marked for dialect, and that he says their speech is purer than that spoken by children in most parts of England. By 'pure' Dixon does not mean that the speech approaches an ideal; rather, he means that the speech is not marked heavily for dialect. All of this is entirely in keeping with what we now know about the development of a local variety of English in a colonial society.

Much the same point as Dixon's was made by George Bennett a decade later in *Wanderings in New South Wales, Batavia, Pedir Coast Singapore, and China* (1834):

> It has often been mentioned by writers upon the United States of America, that a purer and more correct English is spoken in that country than in

the 'old country' where it is corrupted by so many different provincial dialects. The remark respecting the United States of America will equally apply to Australia; for among the native-born Australians (descended from European parents), the English spoken is very pure; and it is easy to recognise a person from *home* or one born in the colony, no matter what class of society, from this circumstance.[17]

The word 'purer' appears again, and again the observer notes the absence of strong dialectal features in the speech of the native-born. Across all classes, the native-born all speak with the same accent—the foundation accent has been established. Because it is the foundation accent it will not be possible for other accents—which will continue to be transported to Australia—to have any significant effect on this foundation accent.

Caroline Leakey's *The Broad Arrow* was published in 1859, but it relates to the years 1848–53 when the Exeter-born Leakey spent five years in Tasmania with her sister. She makes some interesting observations on the language of the Tasmanian-born when describing Hobart:

> The incongruous medley of shops, rich and poor together, is London-like. Butcher, baker, grocer, all appear to have served their apprenticeship in the capital; the cut of the meat, the shape of the bread, the adulteration of the groceries, are in dutiful or unintended remembrance of Cockney education.
>
> Are all the tradespeople of London origin that it should be so, uncle?
>
> By no means. Trades from every part of Britain have settled here. Every county has its representative, every provincial custom its follower. Every grade and every phase of English life meet out here. It is probably this very amalgamation that reproduces the English metropolis.
>
> To the same cause may be attributed the freedom from peculiarity in the tone and pronunciation of the natives. As children they have no opportunity to contract the nasal twang or gutturals of any particular province; by the constant change of servants, and from an intercourse with a diversity of accents, they are preserved from fixing on any one peculiarity. The Irish brogue heard today is to-morrow changed for the broad Scotch accent; the Devonshire drawl soon forgotten in the London affectation; the Somersetshire z's are lost in Yorkshire oo's. If you have not already remarked it, you cannot fail shortly to note how very well the common children speak, even where the parents set them no good pronunciative example.[18]

This passage provides further evidence that the Australian accent developed from a process of the levelling and excision of obvious dialectal features. Leakey was obviously familiar with the social and business milieu of the London

Cockneys, and although she noted their influence on the worlds of butcher, baker and grocer, she heard nothing in the speech of the native-born Australians to remind her of Cockney speech. She says they speak 'well', clearly a synonym for the 'pure' of Dixon and Bennett.

Finally, in 1886, James Froude in *Oceana, or, England and her colonies* comments: 'The first thing that struck me—and the impression remained during all my stay in Australia—was the pure English that was spoken there.'[19]

How, then, can we explain these different evaluations of the Australian accent? Until the mid-1880s, the accent of native-born Australians was regarded as pure, and it was praised for being free of any elements of British dialects. In the first half of the twentieth century, it was regarded as impure, ugly and substandard. It is simply not credible to suggest that the accent changed in one generation. The only explanation possible is that something caused a change in attitude towards the Australian accent.

We are fortunate in having evidence that demonstrates this change of attitude taking place. From the mid-nineteenth century, school inspectors in the various colonies (and later, states) reported annually to parliament on the standards of school students. The typical report dealt with such aspects as reading, writing, composition, spelling, arithmetic, grammar and geography. One very early report from 1855 makes a comment that is potentially in keeping with the twentieth-century comments: 'Little care is apparently taken to correct vicious pronunciation or improper modulations of voice, and we often had occasion to remark, while hearing the children read, that this inattention has a tendency to foster an Australian dialect which bids fair to surpass the American in disagreeableness.'[20] This is not, however, typical, and most of the comments for the next three decades are about non-standard features such as h-dropping, and general observations about carelessness in language use.

Here are some typical comments: from 1863, 'The commoner defects in the articulation are the drawling out of one or more of the vowel sounds, and the substitution of r for that of the terminational w';[21] from 1875–76, 'The omission of the letter *h* (even in the reading of the pupil-teachers) is the chief defect';[22] from 1876–77, 'The chief faults of pronunciation against which a constant struggle has to be carried on are the misplacing of the letter *h*, the omitting to sound *s* at the end of a word, and the withholding its proper force from the final -*ing*'.[23]

From 1879–80:

> There are some words that one hears mispronounced in the schools more frequently than others. For example, many children append in pronunciation a *t* to these words: *cliff, once, sudden*; misplace the accent

in *distribute, executive, laborious, mischievous*; and give the wrong sound to the last syllable in *massacre*.[24]

From 1884–85:

> The children drop their h's without correction, and in some cases the teachers set them the example. From the Cockney and Cornish the disease has spread to the Scotch and Irish. We are in a fair way of becoming a nation without an h in our vocabulary…There are other weak points in Victorian pronunciation; but this, I think, is the worst.[25]

From 1886: 'The most common defects are…disregard of the aspirate, and a corruption of *ing* into *en*, as written, readen for writing, reading.'[26]

The one exception to this early emphasis on non-standard features is an early concern with the pronunciation of the diphthong heard in the word 'cow'. An 1861 report from NSW comments briefly on the pronunciation *caow* for *cow*,[27] and an 1869 Tasmanian report similarly comments on 'cow': one of 'the two gravest defects in the pronunciation of young Tasmanians…[is] the drawling of the diphthongs *ou* and *ow*, as in "maountain" for "mountain", "taown" for "town", and so on'.[28] But such comments are not typical of the early period, and no further comments on the quality of vowels and diphthongs appear until the mid-1880s, with the heaviest emphasis being in the 1890s and the first decade of the twentieth century. In this later period, the sounds commented on most commonly are the diphthongs that occur in 'take', 'hide', 'cow' and 'go', and the vowels that occur in 'hat' and 'get'. A report from 1891–92 comments on 'take', claiming that 'tail' is often pronounced as 'tile'.[29] A Victorian report from 1891–92 comments on 'hide', claiming that 'miners' is often pronounced as 'moiners'.[30]

The early comments on 'cow' in NSW and Tasmania have been noted; an 1891–92 Victorian report takes this up: 'The mispronunciation of…the diphthong "ow", as in brown…is frequently allowed to pass uncorrected. Thus a child might read—"The 'breown' bear".'[31] An 1893–94 report comments on 'go': 'The vowel sounds suffer badly in our schools…go becomes gao.'[32] A report from 1893 focuses on 'hat': 'The short *a* is sometimes pronounced like *e*, e.g., *ketch* for *catch*.'[33] An 1885–86 report comments on 'get', claiming that 'Ben' appears as 'Bin'.[34] These are among the most important sounds that distinguish the Australian accent from other English accents.

For the purposes of this essay, I will focus on the sound in 'take'. This sound is the difference most commonly perceived between Australian English and received pronunciation and other Englishes, and is regarded as the most distinctive of the Australian sounds. The Australian English realisation of the diphthong often results in non-Australian speakers perceiving the sound as closer to 'mike' than 'make'.

Comments on this sound in the school inspectors' reports from the period 1886 to 1914 include: from 1885–86, 'The following errors, I have noticed, are very prevalent…vowel mispronunciation − , ai, ay = ī + (very short); e.g., late = lī't, rain, day, etc.'[35] From 1891–92: 'The mispronunciation of some of the vowel sounds (especially…the long "a", as in tail…is, however, still very common, and is frequently allowed to pass uncorrected. Thus a child might read—"The [brown] bear has a very short 'tile'"…and such pronunciation be unnoticed.'[36]

From 1893: 'The long *a* is often made into a diphthong: I have a lively recollection of hearing the song "Our Jack's Come Home To-day" cheerfully rendered "Our Jack's Come Home to Die!"'[37] From 1893–94: 'I have reluctantly to confess that the vowel sounds suffer badly in our schools—that page becomes pahidge.'[38] From 1899–1900: 'The long *a* sound is too frequently given very much like the long *i* sound, and I am frequently reminded of the lady's invitation to take some cake in preference to grapes—"Tike some kike, I mide it myself; you can have the gripes after".'[39]

Also from 1899–1900: 'The sound of "a", as in fate, is frequently heard both in the reading and oral answers of the children as a diphthong closely approximating to the sound of "i" in *mind*. It is somewhat startling to be told that Limerick is noted for the manufacture of *lice* (lace).'[40]

From 1902–03:

> It is hard to show them the difference between *sail* and *sile* (nearly), *kaite* and *kite*…The long 'a' is the vowel which suffers the greatest distortion. 'Later', heard apart from any context, would be easily mistaken for 'lighter', and 'rain', sounded under like circumstances, is hardly distinguishable from 'Rhine'…Playgoers are familiar with the cry of the itinerant vendors of refreshment between the acts. 'Apples, gripes (grapes), lemonide (lemonade).'[41]

From 1903–04:

> With the vowels, the ah sound, which is the first attacked in teaching deaf mutes to read and speak, is the easiest of all; it necessitates the least exertion of the organs of speech, and on this account the children attempt to form all the vowel sounds with the lips, tongue, throat, &c., in position to make this sound. Hence sah-eel (sile nearly) for sail.[42]

From 1906–07: The vowel sounds are well attended to. *Sy* for *say*, *soide* for *side*…are not frequently heard; but they do still occur, and require constant attention.'[43] From 1908–09: 'The making of the long "a" sound into "i" [is] still rather common in many schools.'[44] From 1910: 'A few vulgarisms of pronunciation not infrequently assail one's ears, such, for instance, as…"pide" for "paid".'[45]

From 1911–12: 'I am glad to say that the vowel sounds, which were rather bad at first, are receiving much attention. The "da-owns" and "ta-owns", and "si-lings daown the by", and "moindings" of the "boikes" are slowly disappearing.'[46]

At the same time as the school inspectors and the writers in the educational journals were commenting on the transformation of 'make' into 'mike', other commentators joined the chorus. In 1887, the Scottish-born Victorian schoolteacher Samuel McBurney collected material on pronunciation in Australia and New Zealand for A. J. Ellis's Early English Text Society books on the pronunciation of English. McBurney believed that some of the distinctive Australian sounds were the same as corresponding sounds in Cockney, although he did not argue that Cockney influenced Australian, and he seemed to use 'Cockney' as a blanket term for 'non-standard'. It is interesting that among the features he notes as being shared by Cockney and Australian are precisely those that appear in the reports of the school inspectors, especially the diphthong in 'take' ('alteration of *a* in *fate*, to nearly *i* in *bite*'), 'cow' ('alteration of the first factor of *ow* in *cow*, so that it is written *kyow*, or *caow*') and 'go' ('alteration of *o* in *hope*, to nearly *ow* in *how*').[47] The distinctive 'take' diphthong was also noted in 1892 by visiting British writer G. L. James:

> As to the English spoken in Australia, I believe it has already been remarked how correct, as a rule, it is, and I think it is free from any distinguishing accent or provincialism to a marvellous extent, while the tone of voice is pleasing and well modulated. In Sydney, however, more particularly the young girls, especially of the lower classes, are apt to affect a twang in pronouncing the letter *a* as if it were *i*, or rather *ai* diphthong.
>
> Thus, the refrain to a well-known song will be pronounced as follows:—
>
> Is this a dream?
> Then wiking would be pine.
> Oh! do not wike me,
> Let me dream agine.[48]

In December 1893, the Chief Justice of Victoria, Sir John Madden, at the speech day of the Methodist Ladies' College in Melbourne, made a number of comments on Australian English. He exhorted young Australians to 'pronounce the English vowels as they were intended', and berated them for mangling their diphthongs: 'The average Australian youth turns a into i, and o into aow or eow. He says "dy" instead of "day", "keow" instead of "cow", and "heow" instead of "how".'[49]

Thus in the late 1880s, and especially in the 1890s and the first decade of the twentieth century, there developed a prescriptivist attitude towards Australian vowels and diphthongs—they were being judged against an 'ideal' or 'standard' pronunciation, and the way Australian English diverged from that was noted increasingly.

Where does this standard come from? There has been much debate about the development of received pronunciation. In the comments on accent in Britain in the eighteenth and nineteenth centuries, it is often difficult to know if the writer is simply recommending the excision of dialectal sounds, or is urging a move towards a commonly accepted, or 'received', ideal. The earlier comments, however, seem concerned largely with excision. Thus in the eighteenth century, when Mrs Elizabeth Montague asserts that provincial accents are not acceptable 'in this polished age' and avers that the 'Kentish dialect is abominable, though not as bad as the Northumberland and some others',[50] there is no suggestion that she has an armoury of ideal vowels and diphthongs to fight off the barbarians.

Lynda Mugglestone has demonstrated that by the end of the eighteenth century there was developing a concept of standard speech, indicated by books such as John Walker's *Critical Pronouncing Dictionary* of 1791.[51] As Honey has shown, however, the widespread dissemination of the standard that was to become received pronunciation was a post-1850 phenomenon, based initially in the public school system.[52] Australian educators became aware of this new standard of correctness, and the comments of the school inspectors and the writers of the educational journals reflected increasingly the growing awareness that Australian vowels and diphthongs differed from this new standard.

Another interesting feature of the Australian school inspectors' reports and the educational journals that start to note 'faulty' vowel sounds is that they reveal a developing interest in elocution in Australia.[53] There is much discussion in the inspectors' reports and in the educational journals of elocution books, elocution teachers and of the need for teachers to retrain in the principles of elocution. Elocution is, of course, concerned primarily with the clear articulation of sounds, rather than with the production of particular 'correct' sounds, but it is not surprising that once the notion of correct sounds develops it becomes linked inextricably to elocution. In the 1890s in Australia, elocution comes to include not just articulation of sounds but the production of correct sounds—or, as the commentators describe them, 'pure' sounds. This is not, however, the purity of Dixon, Bennett and Froude—where purity was the excision of dialectal features. Purity is now an ideal pronunciation, against which the bastard breeds can be judged.

Typical of the new emphasis is the appendix to the Victorian inspectors' reports for 1890–91 by George Lupton, a lecturer in elocution. It summarises his conclusions for the year after giving elocution classes in Ballarat, Geelong, Castlemaine and Melbourne: 'During the past year, attention has been given to the study of expression in reading, also to the purity of vowel pronunciation in vocality.'[54] The 1892–93 Victorian report stresses that the 'purity of the vowel sounds' needs attention.[55] The 1893–94 Queensland report said that 'the most common faults were imperfect articulation of consonants, impure vowel sounds, and lack of voice'.[56] The 1898 *Prospectus* for Wesley College in Melbourne reported that the headmaster, Thomas Palmer, had introduced elocution lessons for the boys in order to maintain 'above all purity of vowel sounds'.[57] A 1901–02 Victorian report asserts that there is 'much to be done' to achieve 'the production of pure vowel sounds'.[58]

This emphasis on the purity of vowels and diphthongs continued with the introduction of phonics in the first decade of the twentieth century (the *Oxford English Dictionary*'s first citation for this method of teaching reading is 1901). The teaching of phonics in NSW was being recommended by 1908 in the educational journals: 'Phonic exercises are necessary when the vocal organs are supple and adjustable, when the child is largely dominated by the influence of imitation. It should be possible to modify, if not eradicate, our so-called "Colonial twang".'[59]

A 1909 issue of the *Public Instruction Gazette* provides teachers with phonics exercises to use in classes.[60] In 1916, the *Education Gazette* provided evidence that the teaching of the correct pronunciation of vowels had become part of the Revised Syllabus.[61] A series of articles in the *Gazette* by Hyacinth M. Symonds explains how students can be taught the correct pronunciations of the vowels and diphthongs in the words *go, day, try, cat, eat, school* and *cow*, including information about correct lip and tongue positions.[62] In Victoria, the new courses in phonics seem to have had the desired effect. The 1911–12 report states: 'The short daily drills in phonics are certainly having a good effect on the speech of our scholars: vowel sounds are becoming purer.'[63] The 1912–13 report states: 'The effect of the training given to teachers by the instructors in elocution is markedly good, especially in the matter of clearer and more correct utterance. The vowel sounds are purer; the enunciation is more careful.'[64]

This material has important consequences for our understanding of the history of the Australian accent. The scholarly study of Australian vowels and diphthongs begins with Mitchell[65] and is cemented by Mitchell and Delbridge[66] and Bernard.[67] More recent studies by Harrington et al.,[68] Cox[69] and Cox and Palethorpe[70] have noted some slight shifts since the 1960s, but these shifts are not sufficiently significant to alter perceptions of those sounds that are most distinctive in Australian English. In 1946, Mitchell proposed a distinction between 'educated Australian' and 'broad Australian',[71] but after their detailed studies of the speech of secondary students from across Australia in the early 1960s, Mitchell and Delbridge proposed three varieties, which they called broad Australian (spoken by 34 per cent of the population), general Australian (55 per cent) and cultivated Australian (11 per cent).[72] There is no evidence for the existence of the cultivated variety in the nineteenth century, but we are now in a better position to understand what generated it: it was the product of the desire to move towards received pronunciation.

There is little evidence for the existence of broad Australian in the nineteenth century either. Research by Felicity Cox and Sallyanne Palethorpe on recordings of 12 rural men and women from central-western NSW and Tasmania, who were born in the 1880s, is revealing some surprising results. The researchers were expecting to find evidence of broad accents, since broad Australian was associated especially with rural areas, but the speakers had general Australian accents.[73] Is it possible that broad Australian is a later development, just as cultivated Australian clearly is?

Central to the argument of this essay has been the fact that the noting of the quality of the vowels and diphthongs does not represent a shift in vowel or diphthong quality, but rather a shift in perceptions. G. L. James claimed he heard the make/mike shift in Sydney among 'particularly the young girls, especially of the lower classes', and Justice Madden described the diphthongs of the 'average Australian youth'. One might be tempted to interpret this as evidence of a broadening among the youth, but the school evidence—in one case going back to 1860—shows that what is new is not the phenomenon itself, but the noting of it. Cox suggests that broad Australian 'could have started in the First World War, and could be something to do with the diggers'.

World War I was certainly important in generating many new Australian words;[74] the colloquial language of ordinary Australians was published in the numerous newspapers and magazines of the soldiers, and part of the diggers' sense of Australian identity was the awareness of how different their accent was to those of the British soldiers. I think broad Australian needs, however, to be seen in a wider context. Just as cultivated Australian is the product of the influence of Britishness and empire—among the 'imperial symbols' of Wellings' argument—so broad Australian becomes associated with 'the vernacular culture of the Australian people'. In the first half of the twentieth century, cultivated and broad Australian sit as extremes on either side of general Australian. The extremes have their cultural manifestations: cultivated Australian quintessentially on the ABC until the mid 1960s, broad Australian in a tradition that begins with the works of C. J. Dennis (thousands of copies of the 'trench' edition of *Songs of the Sentimental Bloke* were distributed to the troops in World War I) and climaxes even as a parody of itself in Afferbeck Lauder's *Let Stalk Strine* (1965) and Barry McKenzie.

By the 1970s, attitudes towards Australian English had changed radically. In 1976, there appeared the first general Australian dictionary, edited in Australia and representing the Australian language. This was Graeme Johnston's *Australian Pocket Oxford Dictionary*. The *Macquarie Dictionary* appeared in 1981. In 1987, the government report *National Policy on Languages* stated: 'Australian English is a dynamic but vital expression of the distinctiveness of Australian culture and an element of national identity.'[75] In 1988, there appeared *The Australian National Dictionary: A Dictionary of Australianisms on Historical Principles*, edited by W. S. Ramson. In 1992, at a conference on the languages of Australia, David Blair argued: 'Australian English is currently a self-confident dialect, reasonably secure in itself, and prepared to set its own standards. It is confident enough to be open to outside cultural influences, as is the society in which it functions.'[76] Australian English was finally naturalised—and, as if as a consequence of this naturalisation, the two extremes, cultivated and broad Australian, were on the decline.

Since the 1970s, cultivated Australian has lost its former status, and such an accent is more likely to be a hindrance than an advantage in contemporary Australia. Broad Australian is rarely heard, except when consciously 'cultivated' for its advertising or iconising functions—perhaps a reflection of the changing demographics of Australian society, perhaps an admission that its opposing partner, cultivated, is now devoid of power. The wheel has turned full circle. Most Australians now speak general Australian, or, more accurately, Australian—something very similar to the foundation accent of the 1820s.

ENDNOTES

[1] Board, Ruby W. 1927, *Australian Pronunciation: A Handbook for the Teaching of English in Australia*, Government Printer, Sydney, p. 6.

[2] Ellis, A. J. 1869, *On Early English Pronunciation*, vol. 1, Trübner, London, p. 23.

[3] Wellings, Ben 2005, 'Crown and Country: empire and nation in Australian nationalism, 1788–1999', *Journal of Australian Colonial History*, 5 (2005), pp. 148–70 at p. 170.

[4] Mallarky, Edith M. 1914, 'Some Means of Training in Speech and other Papers', *Records of the Education Society*, Government Printer, Sydney, p. 8.

[5] Desmond, Valerie 1911, *The Awful Australian*, J. Andrews & Co., Sydney, p. 21.

[6] Churchill, William 1911, *Beach-la-mar, the jargon or trade speech of the western Pacific*, Carnegie Institute of Washington, Washington D.C., p. 14.

[7] Cited in Mitchell, A. G. 1946, *The Pronunciation of English in Australia*, Angus & Robertson, Sydney, p. 63.

[8] Cunningham, Peter 1827, *Two Years in New South Wales*, vol. 2, H. Colburn, London, p. 60.

[9] Meredith, Mrs Charles 1844, Notes and Sketches of New South Wales During a Residence in the Colony from 1839 to 1844, John Murray, London, p. 50.

[10] Horne, R. H. 1859, *Australian facts and prospects: to which is prefixed the author's Australian autobiography*, Smith, Elder, London, refers to the 'nasal twang' (p. 67).

[11] Boldrewood, Rolf 1979, *Robbery Under Arms*, A. Brissenden (ed.), University of Queensland Press, St Lucia, p. 310. The novel was first published in serial form in 1882–83.

[12] Praed, Rosa 1889, *The Romance of a Station: An Australian Story*, Chatto & Windus, London, p. 22.

[13] Praed, Rosa 1893, *Outlaw and Lawmaker*, Chatto & Windus, London, p. 20.

[14] Mitchell, *The Pronunciation of English in Australia*, pp. 5–10, dismisses most of the stereotypes.

[15] Dixon, James 1822, *Narrative of a Voyage to New South Wales and Van Dieman's Land in the Ship 'Skelton' during the year 1820*, John Anderson, Edinburgh, p. 46.

[16] See Maclagan, Margaret A. and Elizabeth Gordon 2004, 'The Story of New Zealand English: What the ONZE Project Tells Us', *Australian Journal of Linguistics*, vol. 24, no. 1, pp. 41–56; Trudgill, Peter

2004, *New-dialect formation: the inevitability of colonial Englishes*, Edinburgh University Press, Edinburgh. For a survey of earlier theories about the development of the Australian accent, see Cochrane, G. R. 1989, 'Origins and Development of the Australian Accent', in Peter Collins and David Blair (eds), *Australian English: the language of a new society*, University of Queensland Press, St Lucia, pp. 176–86.

[17] Bennett, George 1834, *Wanderings in New South Wales, Batavia, Pedir Coast Singapore, and China; being the journal of a naturalist in those countries during 1832, 1833, and 1834*, vol. 1, R. Bentley, London, p. 331.

[18] Leakey, Caroline 1859, *The Broad Arrow*, Richard Bentley, London, pp. 96–7.

[19] Froude, James 1886, *Oceana, or, England and her colonies*, Charles Scribner's Sons, New York, p. 84.

[20] 'Report from School Commissioners on State of Education, 1855', *Journal of the Legislative Council of New South Wales*, Session 1856–57, vol. 1, p. 269.

[21] 'National Education Report, 1863', *Journal of the Legislative Council of New South Wales*, Session 1864, vol. 11, p. 90.

[22] 'Report of the Minister of Public Instruction for 1875–76', *Victorian Parliamentary Papers*, Session 1876, vol. 3, no. 46, p. 57.

[23] 'Report of the Minister of Public Instruction for 1893–94', *Victorian Parliamentary Papers*, Session 1877–78, vol. 2, no. 47, p. 44.

[24] 'Report of the Minister of Public Instruction for 1879–80', *Victorian Parliamentary Papers*, Session 1880–81, vol. 3, no. 31, p. 190.

[25] 'Report of the Minister of Public Instruction for 1884–85', *Victorian Parliamentary Papers*, Session 1885, vol. 4, no. 74, p. 134.

[26] 'Report of the Secretary for Public Instruction for 1886', *Votes and Proceedings of the Legislative Assembly, Queensland*, Session 1887, vol. 3, p. 488.

[27] 'National Education Report for 1861', *Journal of the Legislative Council of New South Wales*, Session 1862, vol. 9, part 1, p. 334.

[28] 'Board of Education Report for 1869', *Journals of the House of Assembly, Tasmania*, Session 1870, vol. 19, no. 18, p. 17.

[29] 'Report of the Minister of Public Instruction for 1891–92', *Victorian Parliamentary Papers*, Session 1892–93, vol. 5, no. 133, p. 43.

[30] 'Report of the Minister of Public Instruction for 1891–92', *Victorian Parliamentary Papers*, Session 1892–93, vol. 5, no. 133, p. 80.

[31] 'Report of the Minister of Public Instruction for 1891–92', *Victorian Parliamentary Papers*, Session 1892–93, vol. 5, no. 133, p. 43.

[32] 'Report of the Minister of Public Instruction for 1893–94', *Victorian Parliamentary Papers*, Session 1894–95, vol. 2, no. 3, p. 41.

[33] Report in the *New South Wales Educational Gazette*, 1 August 1893, p. 45.

[34] 'Report of the Minister of Public Instruction for 1885–86', *Victorian Parliamentary Papers*, Session 1886, vol. 3, no. 89, p. 153.

[35] 'Report of the Minister of Public Instruction for 1885–86', *Victorian Parliamentary Papers*, Session 1886, vol. 3, no. 89, p. 153.

[36] 'Report of the Minister of Public Instruction for 1891–92', *Victorian Parliamentary Papers*, Session 1892–93, vol. 5, no. 133, p. 43.

[37] Report in the *New South Wales Educational Gazette*, 1 August 1893, p. 45.

[38] 'Report of the Minister of Public Instruction for 1893–94', *Victorian Parliamentary Papers*, Session 1894–95, vol. 2, no. 3, p. 41.

[39] 'Report of the Minister of Public Instruction for 1899–1900', *Victorian Parliamentary Papers*, Session 1900, vol. 2, no. 17, p. 61.

[40] 'Report of the Minister of Public Instruction for 1899–1900', *Victorian Parliamentary Papers*, Session 1900, vol. 2, no. 17, p. 61.

[41] 'Report of the Minister of Public Instruction for 1902–03', *Victorian Parliamentary Papers*, Session 1904, vol. 2, no. 1, pp. 35, 52.

[42] 'Report of the Minister of Public Instruction for 1903–04', *Victorian Parliamentary Papers*, Session 1905, vol. 3, no. 1, p. 37.

[43] 'Report of the Minister of Public Instruction for 1906–07', *Victorian Parliamentary Papers*, Session 1908, vol. 1, no. 2, p. 35.

[44] 'Report of the Minister of Public Instruction for 1908–09', *Victorian Parliamentary Papers*, Session 1910, vol. 2, no. 5, p. 33

[45] 'Report of the Director of Education for 1910', *Journals of the House of Assembly, Tasmania*, Session 1911, vol. 65, no. 9, p. 11.

[46] 'Report of the Minister of Public Instruction for 1911–12', *Victorian Parliamentary Papers*, Session 1913–14, vol. 2, no. 6, p. 57.

[47] Cited from a reprinting of a newspaper article by McBurney in the *Press* (New Zealand), 5 October 1887, in Turner, G. W. 1967, 'Samuel McBurney's Newspaper Article on Colonial Pronunciation', *AUMLA*, 27, pp. 82–3.

[48] James, G. L. 1892, *Shall I Try Australia?*, L. Upcott Gill, London, p. 253.

[49] *Argus* (Melbourne), 21 December 1893, p. 4.

[50] Cited from Honey, John 1989, *Does Accent Matter? The Pygmalion Factor*, Faber & Faber, London, p. 23.

[51] Mugglestone, Lynda 2003, *Talking Proper: the rise of accent as social symbol*, Second edition, Oxford University Press, Oxford, esp. pp. 79–80.

[52] Honey, *Does Accent Matter?*, pp. 12–27; also Honey, John 1997, *Language is Power: The Story of Standard English and its Enemies*, Faber & Faber, London, esp. pp. 101–4.

[53] On elocution in Australia, see Damousi, Joy 2007, '"The Filthy American Twang": Elocution, the Advent of American "Talkies", and Australian Cultural Identity', *American Historical Review*, vol. 112, no. 2, pp. 394–416.

[54] 'Report of the Minister of Public Instruction for 1890–91', *Victorian Parliamentary Papers*, Session 1891, vol. 4, no. 73, p. 118.

[55] 'Report of the Minister of Public Instruction for 1892–93', *Victorian Parliamentary Papers*, Session 1893, vol. 2, no. 41, p. 53.

[56] 'Report of the Secretary for Public Instruction for 1894', *Votes and Proceedings of the Legislative Assembly, Queensland*, Session 1895, vol. 2, p. 954.

[57] Reported in Crotty, Martin 2001, *Making the Australian Male: Middle-class Masculinity 1870–1920*, Melbourne University Press, Melbourne, p. 38.

[58] 'Report of the Minister of Public Instruction for 1901–02', *Victorian Parliamentary Papers*, Session 1902–03, vol. 2, no. 28, p. 78.

[59] *Public Instruction Gazette: The Official Gazette of the Public Instruction Department of New South Wales*, 30 April 1908, p. 297.

[60] *Public Instruction Gazette: The Official Gazette of the Public Instruction Department of New South Wales*, 30 September 1909, pp. 308–17.

[61] *The Education Gazette: Official Gazette of the Education Department of New South Wales*, 1 April 1916, p. 80.

[62] Ibid., pp. 79–80.

[63] 'Report of the Minister of Public Instruction for 1911–12', *Victorian Parliamentary Papers*, Session 1913–14, vol. 2, no. 6, p. 62.

[64] 'Report of the Minister of Public Instruction for 1912–13', *Victorian Parliamentary Papers*, Session 1914, vol. 2, no. 1, p. 60.

[65] Mitchell, A. G. 1946, *The Pronunciation of English in Australia*, Angus & Robertson, Sydney.

[66] Mitchell, A. G. and Arthur Delbridge 1965, *The Pronunciation of English in Australia*, Revised edition, Angus & Robertson, Sydney; Mitchell, A. G. and Arthur Delbridge 1965, *The Speech of Australian Adolescents*, Angus & Robertson, Sydney.

[67] Bernard, J. 1967, Some measurement of some sounds of Australian English, Ph.D. Dissertation, University of Sydney; 'Toward the acoustic specification of Australian English', *Zeitschrift fürPhonetik*, vol. 2, no. 3, 1970, pp. 113–28.

[68] Harrington, Jonathan, Felicity Cox and Zoe Evans 1997, 'An acoustic phonetic study of broad, general, and cultivated Australian English vowels', *Australian Journal of Linguistics*, vol. 17, no. 2, pp. 155–84.

[69] Cox, Felicity 1998, 'The Bernard data revisited', *Australian Journal of Linguistics*, vol. 18, no. 1, pp. 29–55.

[70] Cox, Felicity and Sallyanne Palethorpe 2000, 'Vowel change: synchronic and diachronic evidence', in David Blair and Peter Collins (eds), *English in Australia*, John Benjamins, Amsterdam, pp. 17–44.

[71] Mitchell 1946, *The Pronunciation of English in Australia*, pp. 11–18.

[72] Mitchell and Delbridge 1965, *The Pronunciation of English in Australia*, pp. 11–19.

[73] Reported in 'Rack off Hoges, we just don't like the way you speak', *Sydney Morning Herald*, 25 January 2005, p. 1.

[74] See Laugesen, Amanda 2005, *Diggerspeak: The Language of Australians at War*, Oxford University Press, Melbourne.

[75] Lo Bianco, J. 1987, *National Policy on Languages*, Australian Government Publishing Service, Canberra, p. 72.

[76] Blair, David 1993, 'Australian English and Australian identity', in Gerhard Schulz (ed.), *The Languages of Australia*, Australian Academy of the Humanities, Canberra, p. 70.

9. Voice, Power and Modernity

Bruce Johnson

It has been argued that sound is one of the oldest ways of defining, encroaching on and enlarging territorial space, of manifesting power. From the war cries of the ancients, to the howling of the urban mob, one of these sounds has been the human voice. Until the late nineteenth century, however, the radius of vocalised space was limited by the body. From the 1870s, for the first time in human history, this dynamic was utterly transformed by the invention of the sound recording and subsequent related technologies. These technologies permitted a spatial increase that has ultimately become global and, through sound storage systems, a temporal enlargement. The small voice, the local voice, the domestic voice, all clamour for transnational public space, yet they can also be turned against their source through various sound-mixing technologies. With reference to case studies, this chapter will signpost the ambiguous relationship between the voice and power in the modern era.

***

> So the people shouted when the priests blew with the trumpets: and it came to pass, when the people heard the sound of the trumpet, and the people shouted with a great shout, that the wall fell down flat, so that the people went up into the city, every man straight before him, and they took the city.
>
> And they utterly destroyed all that was in the city, both man and woman, young and old, and ox, and sheep, and ass, with the edge of the sword. (The Book of Joshua, Chapter 6, Verses 20–1)

So perished Jericho, reminding us of the great antiquity of the role of the voice in defining territory and identity, of the connection between voice and power. So too when Hannibal faced the Romans in 202 BC in the battle of Zama that ended his campaign. According to Livy, the voice contributed significantly to this defeat:

> There were…factors which seem trivial to recall, but proved of great importance at the time of action. The Roman war-cry was louder and more terrifying because it was in unison, whereas the cries from the Carthaginian side were discordant, coming as they did from a mixed assortment of peoples with a variety of mother tongues.[1]

My argument here is a very broad brush, extrapolating from a range of case studies that I have conducted elsewhere. I want to take a deep historical perspective on the way in which that relationship between voice and power has changed, to contextualise the particular contention here: that is, that the re-emergence of the voice from the late nineteenth century to challenge print as a major site of power is one of the keys to understanding modernity.

It has been argued explicitly and implicitly through, for example, ethno-musicology, that sound is the most ancient, widespread and durable way of defining the territory through which human beings project their individual and collective identity.[2] Among the categories of sound deployed for these purposes, the voice, with its capacities for the most finely discriminated sonic semiotics, has been prominent. In Western Europe, the authority of the oral was challenged by the advent of print and the spread of literacy from the late fifteenth century. The new technology achieved two things. First, it enabled the widespread dissemination of information in standardised form far beyond the radius of the human voice. Scholars all over Europe could read the same things and, equally important, view the same images and diagrams. The other outcome was to create a new marker of class: those who could read and those who could not. As London commerce, with its peripatetic peddlers and their street cries, demonstrated, everyone could shout, but not everyone could write.[3]

While we think of sixteenth-century England as being a richly literary culture, in fact by 1558 only one in five men and one in 20 women could even write their names.[4] The written text defined a new site of literal and cultural capital. Thus, coinciding with the rise of capitalism, to be able to read and write became a new interface of class confrontation. Wheale opens his study of literacy in fifteenth- and sixteenth-century England by noting that when Christopher Marlowe presents the seven deadly sins in *Doctor Faustus*, the plebeian Envy has developed a new resentment not apparent in his medieval predecessors: 'I cannot read, and therefore wish all books were burnt.'[5] This of course is not to make the absurd suggestion that people gradually stopped talking to each other, and the period was oratorically abundant, in theatre, pulpit and education. But there was a progressive decline in oral and aural sophistication as the dominant position gradually assumed by print in the information economy changed the status of sound in general and of the voice in particular.

These changes over several centuries constitute an instructive study in the complex operations of hegemony in and through intellectual and material culture. While the story of the printed text in England begins in the late fifteenth century, like all technologies, there is a considerable 'take-up' time before its impact changed the norms of a society. Kernan dates the arrival of the 'generally accepted view that what is printed is true, or at least truer than any other type of record' to some time in the eighteenth century, relating it to the pervasiveness of print in the form of posters, bills, receipts and newspapers—part of the fabric of everyday life.[6] This raises the question, 'Whose everyday life?' Samuel Johnson's works, including his *Dictionary*, can be taken as confirming the triumph of print as a form of cultural authorisation, the creation of the community of the 'common reader'.[7] But of course literacy in the mid-eighteenth century was by no means universal. If the idea of the common reader defined and validated membership of the community, clearly very large numbers of the population did not enjoy such authorisation. If Johnson contributed crucially to the idea of a community defined through its readerly competencies, his *Dictionary* also created a zone of exclusion. He can be said to have created the English proletariat by creating the proletariat in English. The word had appeared in the 1660s, but the lexicographic authority of Johnson's *Dictionary* formally located the category at the lowest level of society: 'mean, wretched, vile or vulgar.'[8]

This was one of many ways in which the community of the British nation was imagined into existence during the eighteenth century, from a new flag and anthem through to the codification of national pastimes such as cricket.[9] For the purposes of this discussion, I want to foreground the connection between literacy and privileged membership of that community, or, more specifically, what it implies about exclusion. Johnson's work contributed to the definition of this community as a linguistic entity with agreed protocols of literacy. The lexicon embraced by his *Dictionary* was thereby given stability and permanence. It became stationed in the landscape of the imagination as a form of British property. I use the word 'stationed' pointedly, for its associations also with print and something fixed in space, as opposed to the 'mob', a word associated with noise and rootlessness.[10] All that mattered culturally in the idea of the British nation was to be accommodated in, and thus validated by, the community of its common readers. Thus, the playwright for that richly oral forum, the Elizabethan stage, is progressively recuperated as the supreme 'man of letters' from the late seventeenth century: Shakespeare as the pinnacle of English literature.[11]

It is notable, then, that Johnson excluded from the lexicon of written English the diction of the underclasses as 'unworthy of preservation'.[12] The ability to read was already long acknowledged as a privilege in law through the tradition of 'benefit of clergy'. Johnson's exclusions from his lexicography enlarged the community deprived of such benefits. His refusal to admit the diction of the labouring classes to the approved lexicon of Englishness was part of their progressive criminalisation, which went back to the long-standing refusal of the courts to recognise the 'canting' language of the poor.[13] It also stiffened the disenfranchisement of non-literate cultures (within and beyond 'the nation'). The spoken language—the cant—of the illiterate proletariat was regarded increasingly as in itself evidence of criminality.[14] Seen in terms of the historical confrontations between literate and sonic information circuits, this represents a further stage in the politicisation of noise. Deprived of an 'authorised' (that is, written) language, the proletariat must choose between deferential silence and disruptive and seditious noise. Linebaugh lists a succession of legislative measures from the early eighteenth century that were intended to repress and regulate the working classes. One of these was the *Riot Act* of 1715, and it is relevant to this discussion that evidence of its violation was the gathering in public places of a noisy, riotous, tumultuous group of 12 or more people. [15] Those whose position in the political economy was defined by making public noise were disruptive and thus also proclaimed the probability of their criminal tendencies.[16]

This is a further stage in the process by which silence became associated increasingly with moral and intellectual gravitas, as reflected in treatises on appropriate comportment for reading, as well as in the increasing acoustic regimentation of congregations and theatre and concert audiences. The right to impose silence increasingly defined relations of power. The pathologies that defined the disenfranchised brought into being by the capitalist order included not just supposed illiteracy, but their immersion in the alternative communication circuit: sound. They make noise, and in doing so manifest themselves as a threat to a hegemonic textuality. The ruling orders maintain their power through print; the subordinated are identified in networks of orality. I am suggesting that the struggle over the right to make noise is a very useful way of tracing the history of relations of power since the medieval period. The rise of the mob, the urban crowd, the embryonic working class or the proletariat—those who were oppressed under capitalism—is figured as the rise of noise. Wordsworth finds London beautiful when standing back from it in the silence of the morning on Westminster Bridge. But when, in Book Seven of *The Prelude*, he actually enters its streets, he is disturbed and disgusted by the noise of its plebeian herd, the oppressive 'roar' (line 178), 'deafening din' (line 155), 'thickening hubbub' (line 211), the 'uproar of the rabblement' (line 273).[17]

By the nineteenth century, regimes of silence were imposed on factory workers, and the 'silent system' introduced by George Chesterton as a way of degrading convicts in Coldbath Fields Prison in 1834 was universalised in the *Prison Act* of 1865.[18] In colonial Australia, the singing of songs by convict labourers could be penalised by up to 100 debilitating lashes—enough to flay a man.[19] Foucault's work has led to a fixation on scopic regimes, an obsession with surveillance and 'The Gaze'.[20] Strangely little has been made of this complementary but sometimes contesting dynamic: the politicisation of sound and the attempts to silence the subordinated orders.

The capitalist underclasses remained relatively illiterate until the effects of the *Education Act* of 1871 began to filter through to the lower middle classes and women in particular, much to the alarm of custodians of civilised culture.[21] These underclasses were defined by their noisiness, which was also a way of consolingly limiting the radius of their power, compared with the reach of print to an international hegemonic class. Authority resided in and was controlled by the scopic, by print and the eye. The growing threat of 'the mob' was signalled in the French Revolution, and one sign of the nervousness thus instilled was a strengthening of the imagined link between civilisation and the visual rather than the oral as the instrument of authority and power. In Jacob Burckhardt's influential *The Civilisation of the Renaissance in Italy*, first published in 1860, he articulated the well-springs of that civilisation as a primarily visual culture. Significantly, he described the work as 'a vision'.[22]

Interestingly, it was in the same decade that a technological development that was itself a product of the capitalist urban information economy arrived to dismantle the monopoly of print. In 1877, Edison took out the first patent on a sound-recording device.[23] For centuries, print had severed information from sound. The connection was literally rewired with Morse code, a technology generated by the information explosions associated with urbanisation and the internationalisation of capitalism. The proliferation and circulation of commercial data overtook the capacities of the scribal hand, requiring new stenographic technologies. The noise of the typewriter became the new trope of busy-ness, or business, because, like the sound recording, it came into being for stenographic purposes. As with most new technologies, the sound recording's function was quickly re-defined by the market-place and, like the telephone, it re-audialised society. Other similar technologies have followed, including the microphone, radio, movie soundtracks and audio-internet.

These are not simply technological add-ons to society, but transforming interventions, and in particular they transformed power relations in the modern era. The breaking of the link between scribal silence and the movement of commercial information, for example, completely inverted the gender profile of the 'keeper of the secrets': the secretary. You didn't need a good writing hand to use a typewriter. Before the invention of the typewriter, in Europe, women would be taught to read, but for the most part, they could read only what men had written down. In general, women were not required to write professionally and the teaching of writing to women was erratic. Accordingly, their handwriting was notoriously uneven, ungrammatical and illegible. At the same time, there was an enormous pool of unemployed women: that is, women excluded from the economy, and in particular the information economy. New stenographic technologies transformed these politics. In 1870, only 4.5 per cent of stenographers and typists in the United States were women. By 1930, the figure was 95.6 per cent.[24]

These new information technologies thus destabilised the foundations of cultural capital, most fundamentally in democratising access to information by bringing it back to the realm of the acoustic. It should not therefore be surprising that those who had enjoyed various kinds of power were frequently suspicious of these new information circuits. The French government led the way in attempting to regulate ownership of the information stored on sound recordings by drawing it into the discourses of print. Thus, sound recordings constitute

> a special writing, which in the future will undoubtedly be legible to the eyes and is today within everyone's reach as sound; that by virtue of this repetition of imprinted words the literary work penetrates the mind of the listener as it would by means of sight from a book.[25]

Early advertising for record players addressed the anxieties of the genteel classes regarding the product's vulgarity. Male singers spurned the microphone as a sign of weakness and effeminacy, ironically leaving it to women (in Australia at least) to pioneer vocal techniques that defined twentieth-century popular music.[26] Kenneth Slessor's ambiguous response to the 'new woman' who emerged after World War I was summarised in her addiction to the telephone, recordings and the radio.[27] In relation to this argument, in a particularly telling mixed metaphor, he says of her that 'her eyes are full of wireless'.[28]

The image emphasises the new ascendancy of oral networks through the enormous enlargement of the radius of the voice. Print lost its monopoly in the mass dissemination of knowledge. Individuals who could be disempowered and communities that could be atomised by exclusion from a textual culture now literally found public voices. Once sound could be stored and mass mediated, it provided an alternative network with a presence far more intense than print. And, of all sounds, none projected and constructed identities more intensely than the voice. In 1897, the poet Ernst von Wildenbruch recorded his voice on to a cylinder and wrote a poem for the occasion, including the lines:

> The fawning face can deceive the eye, the sound of the voice can never lie;

> Thus it seems to be the phonograph is the soul's true photograph.[29]

Sound technologies thus restored the ancient link between power and the voice, by enabling the mass circulation of the vocal into spaces that had been dominated by print throughout the Enlightenment. This remains such an unpalatable truth to print-based power blocs that it continues to be denied through academic fixations on the 'power of the gaze', and indeed I suggest that the rise of theory itself in late twentieth-century academia is part of the rearguard attempt to deny the cultural democratisation achieved through mass-mediated sound. 'Theory', of course, is derived from the Greek word for spectacle.

With the increasing diversification and sophistication of acoustic technologies, the voice became an instrument of power that outflanked a medium based on the specialised skills of reading, and the ear again challenged the eye as the source of information. The case of radio amplifies the point. Radio came into general public use in anglophone societies in the 1920s, and, by 1931, 50 per cent of urban families in the United States owned a radio. By 1940, radio news had overtaken the press as the primary source of information and 'the first source of political news' in the United States.[30] Its importance to political debate changed the personal profile of the successful politician. During the US elections of 1928, Democrat candidate Al Smith lost to Herbert Hoover, in spite of the fact that, in person, Smith was by far the more visually arresting speaker than his stolid opponent. One reason was radio. Smith

> could not be persuaded to stand still before a radio microphone and the effect of his voice, with its pronounced East Side accent, moving in and out of earshot, was grotesque and his words unintelligible to many in the South and West. By comparison, Hoover, who was a dull speaker, disciplined himself to talk directly into the microphone, have his shyness mistaken for modesty, and give a general impression of Midwestern sobriety.[31]

It is likely that Orson Welles' notorious 1938 broadcast of *War of the Worlds* would not have had the impact it did were it not for the authority in disaster reportage gained from the broadcast of the Hindenburg disaster in the previous year. The power of the medium was recognised by the more far-seeing—or far-hearing—politicians. One of F. D. Roosevelt's first priorities on gaining office was to set up a committee of inquiry into the role of government in the regulation of wireless broadcasting, and he became one of the most effective pioneers of political broadcasting.[32]

Finally, however, let us recognise the ambiguity of the power of the mediated voice. Roosevelt mobilised the energy of Americans against the Great Depression through his celebrated 'fireside chat' broadcasts. It is less easy to applaud the agenda of the great pioneer of twentieth-century vocalised demagoguery, Adolph Hitler. He first became aware of his own political destiny through his voice. He later wrote of how, as an education officer at a *Reichswehr* camp near Augsburg in 1919,

> all at once I was offered the opportunity of speaking before a large audience; and the thing that I had always presumed from pure feeling without knowing it was now corroborated; I could speak...in the course of my lectures I led many hundreds, indeed, thousands, back to their people and fatherland. I 'nationalised' the troops.[33]

What makes Hitler so 'modern' is not simply his oratorical power, but the effectiveness with which he, along with Goebbels, grafted this to possibilities for mass mediation. In the elections of 1932, through technologised mobility—the aircraft—his energetic 'Hitler over Germany' schedule enabled him to deliver speeches personally across Germany.[34] But it was sound technology that was crucial. Through sound-film and the distribution of more than 50,000 recordings of his 'Appeal to the Nation' speech, Hitler flooded the country with that disturbingly electrified and electrifying voice.[35] Supplemented by rented *Lautsprecherwagons*—vans equipped with external loudspeakers—to fill the streets with Nazi speeches and songs,[36] his voice became the basis of his connection with the German people, and would reach an estimated 20 million citizens as acoustic theatre through the new facility of state radio.[37] The careful stage management of these broadcasts explicitly politicised a paradox that crooners such as Bing Crosby had stumbled on: through the radio voice, it was possible to reconcile the mass with the individual, to speak to everyone as though speaking directly to each.[38] No one demonstrated more effectively than Hitler the reuniting of voice and power in the era of modernity.

Later technology introduced further complexities into the relationship between voice and power. In the 1970s, US jazz writer Whitney Balliett observed that recordings 'take on their own life'.[39] He could not have seen at that time just how profoundly digitisation would vindicate him. So to my last point. Ever since what Serge Lacasse called vocal staging through echoes and ventriloquial effects by ancient shamans and priests, the voice without a visible source has been a bearer of disturbing power.[40] Like the ghost under the stage in *Hamlet*, the *'acousmêtre'*, as Chion called it, has been used to unnerve audiences with the sense of a pervasive yet ontologically ambiguous authority.[41] When that sourceless voice can be mass mediated through digital technology, it becomes potentially a site of autonomous power, power independent of the voice's owner. What von Wildenbruch called the guarantee of self, the 'photograph of the soul', suddenly becomes capable of violating the soul. Thus, towards the end of the twentieth century, the remorseless rise of an Australian demagogue was brought to a halt partly by a vocal doppelganger.

The arrival of Pauline Hanson on the political scene disclosed previously neglected dynamics in Australian society, and she became such a powerful conductor for those disaffected by the move towards globalisation and cultural diversity that John Howard had to delay calling a federal election until the Coalition had absorbed the implications of her stunning success in the Queensland state elections of 1996. But, by 1998, the momentum had gone, and in that year Hanson lost to Cameron Thompson in the bid for the seat of Blair. Thompson attributed his victory in part to the 'battle' Hanson was fighting with her own alter ego, Pauline Pantsdown, during the campaign. Pantsdown was the persona inhabited by Simon Hunt, a lecturer in media at Sydney's College of Fine Arts, and a gay rights activist. I have published a detailed analysis of this episode elsewhere, but refer briefly to it here as a concluding case study in the radical subversiveness of the recorded voice.[42] Pantsdown's most effective weapon in his confrontation with Hanson was the latter's own voice. This was first heard in the dance track *I'm a Back Door Man*, which was constructed out of digitised bites of speeches by Pauline Hanson, laboriously reassembled to present this rabid conservative proudly declaring that she was a lesbian committed to the establishment of a homosexual society.

The idea came to Hunt as he listened to Hanson's voice, with its distinctive cadences and timbres. As we listen to the song and its follow-up, *I Don't Like It*, we are again confronted with von Wildenbruch's axiom. The power of what is spoken lies as much in the voice as in the words, in what is heard as much as in what is understood. What someone says is perhaps of secondary importance to the vocal presence: whose flesh does not tingle at the sound of Hitler's voice, even if we don't speak German? Thus, the development of the sound recording on the cusp of modernity was not simply another way of circulating what someone said. It confirmed modernity as an era of neo-orality, and marked the mass deployment of the power of the voice in a way that finally outreached print.

ENDNOTES

[1] Livy 1965, The War With Hannibal, translated by Aubrey de Selincourt, Penguin, Harmondsworth, pp. 661–2.

[2] See, for example, Schafer, J. Murray 1977, *The Tuning of the World*, McClelland and Stewart, Toronto.

[3] See further Smith, Bruce 1999, *The Acoustic World of Early Modern England*, University of Chicago Press, Chicago and London, pp. 52–71.

[4] Wheale, Nigel 1999, *Writing and Society: Literacy, print and politics in Britain 1590–1660*, Routledge, London and New York, p. 41.

[5] Ibid., p. 1.

[6] Kernan, Alvin 1989, *Samuel Johnson and the Impact of Print*, Princeton University Press, Princeton, p. 49.

[7] Ibid., p. 240.

[8] See the discussion by Linebaugh, Peter 2006, *The London Hanged: Crime and Civil Society in the Eighteenth Century*, Verso, London and New York, pp. 121–2.

[9] Ibid., p. 117.

[10] The two words 'stationery' and 'stationary' converge in the fact that printers could not conduct their business as street pedlars, but required fixed 'stations' of business. The word 'mob' is cognate with 'mobile', being on the move and, as such, a threat to fixed property.

[11] I have developed this and other arguments cognate with this overall discussion, in Johnson, Bruce, 2006, 'Divided loyalties: literary responses to the rise of oral authority in the modern era', *Textus*, Vol. XIX, pp. 285–304.

[12] Linebaugh, *The London Hanged*, p. 429.

[13] Ibid., pp. 71–2.

[14] See, for example, ibid., p. 72.

[15] Ibid., p. 17.

[16] Ibid., p. 144; for the nineteenth century, see Picker, John M. 2003, *Victorian Soundscapes*, Oxford University Press, Oxford and New York, pp. 41–81.

[17] See further Johnson, Bruce 2003b, 'Writing noise: noisy writing: "the eyes no longer have to do their work"', *Proceedings of the International Conference of the World Forum for Acoustic Ecology, Melbourne, March 19–23 2003*, CD Publication, edited by the International Committee of the World Forum of Acoustic Ecology under the Chairmanship of Nigel Frayne, published by the Australian Forum for Acoustic Ecology and the Victorian College of the Arts, Melbourne, 2003, unpaginated.

[18] Babington, Anthony 1971, *The English Bastille: A History of Newgate Gaol and Prison Conditions in Britain 1188–1902*, Macdonald, London, pp. 190–3, 221–2.

[19] Hughes, Robert 1987, *The Fatal Shore: a history of the transportation of convicts to Australia 1787–1868*, Collins Harvill, London, p. 480.

[20] See, for example, Foucault, Michel 1977, *Discipline and Punish: The Birth of the Prison*, translated by Alan Sheridan, Allen Lane, London.

[21] Carey, John 1992, *The Intellectuals and the Masses: Pride and Prejudice Among the Literary Intelligentsia, 1880–1939*, Faber & Faber, London and Boston, see for example, pp. 6, 58.

[22] In a letter, he described historiographical knowledge in terms of visual tropes: 'My starting point has to be a vision, otherwise I cannot do anything. Vision I call not only optical, but also spiritual realization; for instance, historical vision issuing from the old sources.' (Cited by L. Goldscheider in his 'Foreword' to Jacob Burckhardt 1955, *The Civilisation of the Renaissance in Italy: An Essay*, Phaidon Press Ltd, Fifth Printing of the 1944 edition based on S. G. C. Middleton's translation of 1878, p. x.)

[23] Gelatt, Roland 1977, *The Fabulous Phonograph: 1877–1977*, Second Revised Edition, Collier, New York, p. 21.

[24] Kittler, Friedrich 1999, *Gramophone, Film, Typewriter*, translated by Geoffrey Winthrop-Young and Michael Wutz, Stanford University Press, Stanford, California, originally published in German, 1986, p. 184.

[25] Attali, Jacques 1985, *Noise: The Political Economy of Music*, translated by Brian Massumi, Manchester University Press, Manchester, p. 98.

[26] Johnson, Bruce 2000, *The Inaudible Music: Jazz, Gender and Australian Modernity*, Currency, Sydney, pp. 81–105.

[27] Slessor, Kenneth 1981, 'The Voice on the Wire', p. 12, and 'The Green Rolls Royce', p. 32, in *Darlinghurst Nights*, Angus and Robertson, Australia; Slessor, Kenneth 1983, 'Underwood Ann', p. 10, 'Great-grandmother speaks', p. 19, and 'Silence', p. 25, in *Backless Betty from Bondi*, Angus and Robertson, Australia.

[28] Slessor, *Backless Betty from Bondi*, p. 32.

[29] Kittler, *Gramophone, Film, Typewriter*, p. 79.

[30] Cashman, Sean Dennis 1989, *America in the Twenties and Thirties: The Olympian Age of Franklin Delano Roosevelt*, New York University Press, New York and London, pp. 320, 334, 336.

[31] Ibid., p. 107.

[32] Ibid., pp. 325, 336.

[33] Kershaw, Ian 2001, *Hitler*, two volumes, Penguin, London, first published 1998, Vol. 1, p. 124.

[34] Ibid., pp. 363, 364, 369.

[35] Ibid., p. 369.

[36] For this information and a general account of the importance of sound in the emergence of the Nazi Party, see Birdsall, Carolyn (forthcoming), *'Affirmative Resonances' in the City? Sound, Imagination and Urban Space in Early 1930s Germany*. Thanks to the author for permission to read and refer to a pre-publication draft.

[37] Ibid., pp. 433, 440, 453.

[38] See further, Johnson, *The Inaudible Music*, p. 101.

[39] Balliett, Whitney 2000, *Collected Works: A Journal of Jazz 1954–2000*, St Martin's Press, New York, p. 563.

[40] Lacasse, Serge 2000, Listen to my voice: the evocative power of vocal staging in recorded rock music and other forms of vocal expression, PhD, University of Liverpool. See also Connor, Steven 2000, *Dumbstruck: A Cultural History of Ventriloquism*, Oxford University Press, Oxford.

[41] Chion, Michel 1994, *Audio-Vision: Sound on Screen*, translated by Claudia Gorbman, Columbia University Press, New York, pp. 71–3, 129–131; see further Johnson, Bruce 2005, '*Hamlet*: voice, music, sound', *Popular Music*, Vol. 24, No. 2, pp. 257–67.

[42] Johnson, Bruce 2003a, 'Two Paulines, two nations: a case study in the intersection of popular music and politics', *Popular Music and Society*, Vol. 26, No. 1, pp. 53–72.

10. Modernity, Intimacy and Early Australian Commercial Radio

Bridget Griffen-Foley

On its relaunch on 23 March 1936, the Sydney *Daily Telegraph* declared itself 'thoroughly modern'—as modern as television, wireless and airmail.[1] While it was to be another two decades before television arrived in Australia, broadcasting, in the form of radio, was already an entrenched part of the Australian media and entertainment industries. Marshall Berman asserts that the maelstrom of modern life has been fed from many sources: great discoveries in the physical sciences, the industrialisation of production, immense demographic upheavals, rapid urban growth, systems of mass communication, increasingly powerful national states and bureaucracies, and mass social movements of people.[2] In this schema, and in the views of the radio and advertising industries, radio—or, to use the more evocative term, *wireless*—was the quintessence of modernity. Its ability to send through the ether signals and pulses, music and words, banter and instruction seemed a marvel of science and, in the 1920s, the radio tower, like the skyscraper, began to dot the skylines of Australian cities.

As a director of the Australian arm of the powerful advertising agency J. Walter Thompson, W. A. McNair, observed in a pioneering study in 1937, radio was truly established as 'a mass medium'.[3] But, as the American cultural historian Roland Marchand has noted, 'radio surpassed all others in its capacity to deny its own status as a *mass* medium'.[4] This chapter considers how the Australian radio industry welcomed radio's growing reach and economic power while exploiting its potential for intimate contact with its audience. This chapter is, in other words, about how early Australian radio blended the rhetoric of modernity with compensatory varieties of 'personal' contact.

From 1923, when broadcasting began officially in Australia after a series of fits and starts, wireless was addressed in the same terms of scientific wonder as had greeted moving pictures and the gramophone. The modernist magazine *Home* predicted that the technology would 'creep into every phase of your existence. It is the Radio Age.'[5] In an editorial entitled 'The March of Science' in 1927, Sydney's leading radio periodical, *Wireless Weekly*, traced mankind's efforts to gather and disseminate knowledge from the Greek philosophers, physicists and physicians, to scientists during the Renaissance and the wonders of industrialisation: the steam engine, the telescope and the electric telegraph. 'No scientific achievement, however, seized upon popular imagination' like wireless, declared the editorial: 'The profound mystery of ethereal voices and messages provoked the interest of a vast mass of people.'[6]

For years, expectations of radio had been particularly high in Australia, and it was not just *Wireless Weekly* that believed that the medium could, and would, prove to be 'a power for good'.[7] Sir Ernest Fisk, managing director of Amalgamated Wireless (Australasia), enthused that no 'scientific discovery offers such great possibility for binding together the parts of our far-flung Empire'. Former Prime Minister W. M. Hughes asserted that by radio's 'magic the vast globe shrinks to compassable dimensions, distance is annihilated, isolation banished...Wireless will make [imperial ties] still stronger'.[8] Radio, it was confidently thought, would draw the mother country and the dominions into the imperial family, the citizens of the city and the outback into the national family, and members of the household into a family circle.[9]

In 1925, *Wireless Weekly* insisted that Australia's geography should help, rather than hinder, radio's development as 'those people separated by distance from centres of amusement and education must directly benefit'. Indeed, radio, like aviation, would help to overcome the problems of a sparse population defending such a vast area. The publication which did so much to portray wireless as a symbol of modernity and progress and a consumer necessity believed that radio would help to bring 'city joys to country lives'.[10]

In the late 1920s, as the regulatory regime stabilised and the number of radio licences taken out by households grew, there was a discernable shift in the style of presentation on Australian radio. For many early listeners, the great fascination of radio had been in its outside broadcasting as it brought the wider world—big sporting contests, political gatherings and the exploits of aviators—into the privacy and comfort of listeners' homes. But, as Lesley Johnson shows, by the end of the decade industry leaders and periodicals were convinced that radio was most successful when intimate, human and personal. If radio had begun by exploiting the extraordinary, by the 1930s, it was conjuring up a world it shared with listeners—the everyday, ordinary, intimate world of home life.[11] Many individual station slogans elected to draw on notions of companionship rather than on regional characteristics: 2HD Newcastle was 'The Voice of Friendship', 2MW Murwillumbah 'The Friendly Station', 4WK Warwick 'The Listener's Companion', 6AM Northam 'The Happy Station' and 6PM Perth 'The Cheery Station'.[12]

As early as 1926, the radio 'Uncle' and 'Aunt' were a feature of several ABC and commercial stations. Adopting a position of benign superiority, these announcers appealed to adults as well as children. One listener wrote to a radio station in 1928: 'I have a very old Aunt...When the children's hour comes on, I place the loud speaker beside her...I caught her the other day using it as a telephone to talk to the aunts and uncles.'[13] Although women performed in dramas and comedies, most were confined to the role of Aunt in sessions aimed at women and children. Radio was geared to the rhythms dictated by work, housework and school: in the morning, the whole family was included; by 10.30 or 11 am, programs were designed for housewives; sessions that began about 5 pm were tailored to children; and, from about 6.30 pm, came music, variety shows and serials. An Uncle or Aunt might reappear to read a bedtime story.[14] John Dunne, studio manager of 2SM Sydney and leader of 'Uncle Tom's Gang', came to dislike how most Australian stations followed the BBC's example of closing down with a bald 'goodnight'. He introduced a nocturnal benediction for 2SM listeners: 'And so to bed. Sleep in peace, wake in joy, good angels guard thee. And goodnight. God bless you.'[15]

Refining their populist styles during the Great Depression, radio's Uncles and Aunts carefully ministered to their audiences. The first edition of the *Listener In* for 1930 featured a photograph of a pretty, smiling young woman with the caption, 'Start 1930 with a Smile'. Readers were advised to pull together cheerfully and help each other in the face of the economic malady, and to use radio to 'tune out the gloom'.[16] In her 1930 Christmas message, 'Auntie Goodie' (Goodie Reeves) of 2GB Sydney urged her nieces and nephews to 'specially appreciate your presents this year, because most Mummies and Daddies have not had too many pennies to spend'.[17]

'I like you—I'm your friend!' declared the 'Cheerio Man' (Captain A. C. C. Stevens) on a promotional postcard issued by 2UE Sydney in the mid-1930s.[18] The providential ambiguity of the second-person pronoun in English—the 'you' that embodies the intimate, singular *tu* and the impersonal, plural *vous*—allowed the broadcaster to use a form of address that was both mass delivered and highly personal.[19] 'Elva' wrote to *Wireless Weekly* in 1931 to praise the 'personal touch' of radio; her favourite announcers all had 'a friendly, cheerful way of making one feel that in many ways a friend is chatting'.[20]

The successful announcers of the 1930s exploited the technology of radio to create a world of the every day and the ordinary and to produce personalities for themselves 'just like you and me'. The most popular male announcer in Victoria, Norman Banks, set himself up as a friendly advisor in programs such as *Help Thy Neighbour* and *Husbands and Wives*.[21] One broadcaster who specialised in presenting advice programs was Sydney's Frank Sturge Harty, whose appeal to the individual through a mass medium was captured by the *Wireless Weekly*'s label, 'The Man Who Knows the Confidence of TEN THOUSAND WOMEN'. Interviewed about his new 2UE advice program, *Between Ourselves*, in 1938, Harty asserted that it was 'designed to be intimate, so that every woman, sitting alone, will feel that it belongs to her individually'.[22] Comedy programs and soap operas with titles such as *Mr and Mrs Everybody* (networked from 2CH Sydney) and *Mrs 'Arris and Mrs 'Iggs* (networked from 2UW Sydney) conveyed a world familiar to listeners.

Before the advent of talkback radio in the 1960s, there were many ways for Australian listeners to be drawn directly into the world of radio programs and to engage with on-air personalities. Listeners were encouraged to ring in and talk to a studio operator or presenter off air, and to have their stories, jokes or musical requests relayed on air. Commercial radio, in particular, drew on music-hall traditions of audience participation in Community Singing broadcasts, which were a popular feature of radio stations around the country, and programs such as 2UW's enormously successful *Australia's Amateur Hour*.[23] Uncles and Aunts sent birthday calls to individual children, with John Dunne reasoning: 'Call a youngster and that kiddy remains a listener for life.'[24] Radio announcers took their shows on the road and presided over excursions and tours. By the mid-1930s, the private lives of station 'personalities' were very public, with periodicals such as *Radio Pictorial of Australia* reporting on announcers' interests, idiosyncrasies, families, working days and office romances. The radio bachelor even became an identifiable figure in the pages of radio periodicals, with the implication being that you, too, could snare yourself an announcer.[25]

In 1928, the 2UW humorist, Jack Win, noted that each year thousands of letters were received by each station.[26] As one American commentator remarked, radio programs made 'thousands of people feel free to sit down and write a friendly and personal letter to a large corporation'.[27] By the mid-1930s, Australian radio periodicals were gleefully running photos of piles of fan mail, which seemed to attest to the medium's popularity and importance. Station managers and announcers liked them, too. Each day an employee was required to sift through the mail and scrutinise every line in case the letter was in response to a competition. Ardent notes of appreciation found their way into folders to show prospective advertisers. As *Wireless Weekly* put it, '[N]o mail, no proof for manager and advertiser, no job.'[28]

The letters received by radio stations—from children, housewives, invalids and the elderly—were variously sad, amusing, affectionate, business-like, critical or libellous. Some of the signed letters yielded replies, either on or off air. A number of listeners appear to have been quite content to write to their favourite radio personalities confiding intimate details of their lives, including the meanderings of husbands and the misdoings of offspring, with no expectation of a reply. Letters could take serial form, or be accompanied by photographs or gifts. At 2UW, 'Uncle Jack' (J. M. Prentice) received innumerable queries, all of which he tried to answer, no matter how bizarre, as well as compliments, requests and invitations to dances, dinners and tête-à-têtes. By 1936, he had presided over the funeral of one listener and been in touch with another for more than a decade. 'It has helped through many a minor crisis to know that such faithful friendship has been evoked,' he opined. 'Were it not for the friendly word, the appreciation of the fact that an announcer has to work long hours, under great strain, that he must always be cheerful…the job of announcing would be likely to develop into a nightmare.'[29]

Radio stations and networks worked hard to turn their personalities and their endeavours into popular social movements. That 2CH had a 'Fairy Godmother' (Margaret Herd) suggests that the radio family performed benevolent, as well as social and promotional, functions. Radio was represented as a means of escape and fun, of solace and support. The Depression was rendered part of the outside world, with radio clubs emphasising private charitable endeavour and neighbourly help in a depoliticised, personalised realm.[30] Although there were differences between how the 'star' system developed in the film and radio industries, there were also some clear synergies between the social and commercial endeavours of the two most public media of popular culture and communication in the inter-war years. Film and radio interests set up clubs for fans in order to enhance the industries' civic reputability, insinuate themselves in the lives of consumers, engender goodwill and facilitate tie-ins with businesses. From the late 1920s, just before the anticipated arrival of talking pictures, American film interests set up clubs through local radio stations: the MGM Radio Movie Club was established in April 1928 through 2GB; 2UW opened the Fox Movietone Radio Club in 1930; and the Fox Hoyts Radio Club in Perth had evolved into the 6ML Cheerio Club by the mid-1930s.[31]

Probably the most notable of the clubs launched in the early years of Australian radio was the 2GB Happiness Club. The founder was Mrs W. J. (Eunice) Stelzer, a music teacher who had been involved in wartime fundraising. She began performing on 2GB, where her husband worked, shortly after its formation, and joined the permanent staff in 1929. Her women's session drew letters from listeners confiding their worries and seeking advice. As the volume of mail increased, Stelzer organised suburban branches to assist with replies. In September 1929, she announced her intention to form a club to gather together the branches.[32] Two hundred and fifty women attended the meeting at the station's headquarters that spawned the 2GB Happiness Club.

The club's motto was 'Others first', its signature verse *Pull Together*: 'We are helped by helping others; if we give we always get.' Stelzer's brand of self-help, selflessness and sisterhood struck a chord with listeners during the bitter early days of the Depression. The club was non-political and non-sectarian and free to join. Wearing their membership badges, women chatted about the club on the train. Sixty-four branches, each officially opened by Stelzer on 2GB and featuring its own colours and social secretary, were formed across Sydney and beyond.[33]

Like the members of film clubs, Stelzer's ladies had enormous fun doing good.[34] Afternoon teas, 'musicales' and 'conversaziones' were held at David Jones and Mark Foys. Monthly branch meetings usually began with *Advance Australia Fair* and featured singing and dancing. Sometimes 2GB personalities such as 'Uncle Frank' (Frank Grose), host of *Cheer-Up* and *Gloom Chasers* sessions, along with the *Radio Sunday School*, put in an appearance. When one member was thrown a birthday party in 1930, she recorded simply: 'I'll never forget the Genuineness.' The club formed 'Younger Sets' and a vocational training arm for boys, organising aeronautical, electrical and life-saving classes, and tours of manufacturing and engineering works.[35]

Stelzer became the darling of Australian radio periodicals, which delighted in celebrating radio's good works. Her neat dark perm, matronly figure and floral frocks were a familiar sight to readers. In 1932, with membership of the club at 4000 and still growing, 2GB decided to formalise its charitable activities. It established a Charity Board, chaired by G. A. Saunders ('Uncle George'), which included Stelzer, 2GB's managing director, A. E. Bennett, head of the Bluebirds Club for children, Muriel Valli ('Auntie Val'), and Dorothy Jordan. Proceeds from parties, concerts and Community Singing were to be donated to the Dalwood Health Home for Children and, in time, many other institutions. Members of the 2GB Happiness Club visited the sick, repaired shoes and donated and collected clothes for the needy.[36] Its origins in working-class Campsie, its spread to areas such as Wollongong and Woy Woy, and the range of its activities suggest that the club was not directed at a purely middle-class audience, preferring, instead, to speak as if all women were one and the same.[37] And at the centre of this community was 2GB.

While the 2GB Happiness Club formed branches in Exeter and New Zealand, and made arrangements for members travelling to Ceylon to be met at the dock,[38] other clubs proliferated, if on a slightly less ambitious scale. Six months after its formation in 1933, the 2GF Smile Club in Grafton had 1800 members, all of whom pledged to smile three times, and make someone smile, every day. They were given badges and membership certificates showing that 'Uncle Col' (Charles Coldwell-Smith) was president and 'The Smileman' (AWA executive Roger Fair) was patron. Members had their own session on Saturday mornings and, through 2GF, donated flowers and books to the Grafton Benevolent Home. In 1934, Fair introduced the Smile Club to two other AWA stations, 2AY Albury and 3BO Bendigo.[39] The organisers of established clubs were often persuaded to start up clubs elsewhere: in 1939, for instance, Betty Errington went from 3BO to Cairns to establish a women's club, and host the women's and children's sessions, at 4CA. The club rooms at the RSL were surely more salubrious than the studios located in a 'funny, little brown house on stilts' lacking toilet and fan.[40]

Most of the clubs that spread around Australia were christened with upbeat names: the Cheerio Club, the Friendship Circle, the Joy Club and the Look Up and Laugh Club.[41] Feeling that the field was becoming a trifle crowded, Jack Davey formed the Miserable Club in 1934. His first news session facetiously reported on suicides, murders and funerals, and announced the launch of the Back to Long Bay Jail Week and a new serial, *How to Murder in Your Own Home*.[42]

But still the clubs came, including the 5DN Kipling Boys' Club (in Adelaide), the 3SH Women's Club (in Swan Hill) and the 7LA Women's Association (in Launceston). Founded in 1935 by Gordon Marsh, general manager of Hume Broadcasters, the Kipling Boys' Club adopted as its creed Rudyard Kipling's poem *If*. Don Bradman was recruited as president and helped to persuade Kipling himself to be club patron. Under the direction of 'Uncle Bert' (Bert Woolley), 18 000 members worked to raise money for stoves, pianos, prams, clothing, wheelchairs and radio sets for the needy.[43] The 3SH Women's Club was formed in Swan Hill in 1937 to 'promote a bond of friendship' between members, who were approved by the 'senior lady announcer' and selected their own club name. The club hosted debutante balls and a 'Microphone Ball', entered a float in the annual street procession and supported the local hospital.[44] Established in Launceston in 1938, the 7LA Women's Association promoted the station's Community Singing and dances, hosted afternoon teas in its club rooms, raised money and goods for hospitals and orphanages, visited hospital patients and distributed canteen orders to disadvantaged children. Members also joined in activities with the 7HO Women's Association.[45]

By 1939, there were at least 117 clubs affiliated with Australia's commercial radio stations. A list compiled by the Federation of Australian Radio Broadcasters for the Parliamentary Standing Committee on Broadcasting in 1942 indicated that about 40 stations had established women's clubs, with a total membership of nearly 150 000. The committee concluded that women's and children's clubs helped to 'guard morality on the air' and provided a valuable service to the Australian community. There were clubs designed for listeners bound together by age (such as the Call to Youth Club of 2UE, the Junior Country Service Club of 2GZ Orange and the Chatterbox Corner Chums of 3AW Melbourne), interest (the Fisherman's Club of 3XY Melbourne, the Commander Singing Club of 4TO Townsville and the Gardening League of 7EX Launceston), and timeslot (the Night Owls' Club of 2CA Canberra and the Breakfast Clubs of 3AW Melbourne, 3UL Warragul and 3YB Warrnambool), as well as gender.[46]

Like fan mail, radio clubs had commercial overtones and implications. In 1940, an executive with Western Stores, which advertised extensively on 2GZ, declared radio clubs were the 'most useful sessions of any…Men's and women's apparel, domestic appliances, furniture, cosmetics, and foodstuffs can all be successfully sold by utilising the intimate appeal of these clubs'.[47] The Macquarie network catalogues of the early 1940s emphasised each station's 'personalities' and their selling power. The 1942 catalogue boasted that Stelzer and the 2GB Happiness Club offered an audience of tens of thousands of potential customers, while the morning program associated with the 2WL Friendship Club in Wollongong was said to provide enormous 'pulling power for all products!'.[48]

By now there were even clubs for listeners' pets in Melbourne. 3UZ had the Tailwaggers' Pet Club. 3AW's Birmacley Pet Club, sponsored by a margarine manufacturer, had its own session on Saturday mornings and its own membership certificate: 'I solemnly promise…to be faithful to my mistress or master…not to chew slippers or destroy anything in the garden.' The club was headed by Nicky and Nancy Lee, two of Australian radio's biggest stars.[49]

In 1932, Clifford Nicholls Whitta ('Nicky'), a musician who had been working with 3LO in Melbourne, joined 3AW and teamed with Fred Tupper ('Tuppy') to host the breakfast show. The next year, Kathleen Lindgren, who sang and played the ukulele, was brought in to host the children's show. Someone at the station suggested the 'Chums' who listened to *Chatterbox Corner* write in with suggestions for a stage name, and 'Nancy Lee' was born. Nicky hosted *Chatterbox Corner* with Lee, starring as a loveable, naughty boy the Chums and their mums took to their hearts. With its theme song, *Being a Chum is Fun*, it was regarded widely as the best children's show in Australia. Nicky, Tuppy and Nancy Lee hosted fancy-dress and children's parties, arriving in a monogrammed aeroplane, and conducted Christmas and hospital appeals. Their endeavours were hugely successful with audiences and advertisers, and inspired the Chums' Club on another Macquarie station, 7QT Queenstown.

The public fascination with Nicky and Nancy Lee was heightened by their marriage in 1935. When Nancy Lee withdrew from radio to care for their children in 1944, and Nicky decided to leave 3AW in 1946, there was an outcry from listeners: '[W]e feel that we are losing a very close, personal friend,' wrote one. Nicky went on to work for 2CH, 3KZ Melbourne and 3UZ, where he mentored the young Graham Kennedy. Sincere and personal behind the microphone, Nicky gave the impression that he was addressing himself 'to that one person out there whom he liked, understood and felt for'. More than 100 000 people watched his funeral procession through the streets of Melbourne in 1956. When the *Argus* asked members of the great Victorian 'family' of Nicky and Nancy Lee to write in with their special memories, hundreds obliged. At the launch of Nancy's memoir, *Being a Chum was Fun*, in 1979, many former Chums turned up wearing their old membership badges.

At the same time as the Australian radio industry presented itself as the embodiment of modernity, it sought consciously to offset the complexity and impersonality of modern life, fusing its mass appeal with apparent intimacy.

With overseas research indicating that people wanted news and entertainment conveyed through the medium of personalities,[50] Australian radio deployed sophisticated devices to engage the audience and provide compensatory varieties of 'personal' contact. Its personalities strove for a warm, intimate, friendly tone as they addressed their audience as individuals or in small family groups, and tried to approximate a one-to-one relationship. They also reached out to their listeners by considering their problems, leading them in Community Singing concerts, establishing social and charitable clubs and appearing at special events. A modern mass medium—with its resonances of the inchoate, the mechanised and the impersonal—self-consciously transformed itself into a unifying, intimate and highly personalised mass medium, and a cacophony of sounds formed the basis of popular social movements.

ENDNOTES

[1] *Daily Telegraph*, 23 March 1936, p. 8.

[2] Berman, Marshall 1982, *All That is Solid Melts into Air: The Experience of Modernity*, Simon & Schuster, New York, p. 16.

[3] McNair, W. A. 1937, *Radio Advertising in Australia*, Angus & Robertson, Sydney, p. 399.

[4] Marchand, Roland 1985, *Advertising the American Dream: Making Way for Modernity, 1920–1940*, University of California Press, Berkeley and Los Angeles, p. 8.

[5] See Matthews, Jill Julius 2005, *Dance Hall and Picture Palace: Sydney's Romance with Modernity*, Currency Press, Sydney, pp. 114–15.

[6] *Wireless Weekly* (*WW*), 20 May 1927, p. 2.

[7] *WW*, 18 September 1925, p. 6.

[8] Goot, Murray 1981, 'Sir Ernest Thomas Fisk', in Bede Nairn and Geoffrey Serle (eds), *Australian Dictionary of Biography* (*ADB*), Vol. 8, Melbourne University Press, Melbourne, p. 509; Hughes, W. M. 1925, 'Foreward', in J. M. Myers, *Wireless in Australia*, Amalgamated Wireless Australasia, Sydney.

[9] Counihan, Mick 1982, 'The formation of a broadcasting audience: Australian radio in the Twenties', *Meanjin*, Vol. 41, June, p. 201.

[10] *WW*, 8 May 1925, p. 6; 2 October 1925, p. 29.

[11] Johnson, Lesley 1983a, 'The intimate voice of Australian radio', *Historical Journal of Film, Radio and Television*, Vol. 3, No. 1, pp. 43–7.

[12] *Broadcasting Business Year Book*, 1939, p. 100.

[13] Johnson, 'The intimate voice of Australian radio', p. 45; *WW*, 27 January 1928, p. 5.

[14] Shoesmith, Brian and Edmonds, Leigh 2003, 'Making culture out of the air', in Geoffrey Bolton, Richard Rossiter and Jan Ryan (eds), *Farewell Cinderella: Creating Arts and Identity in Western Australia*, University of Western Australia Press, Perth, pp. 220–1.

[15] *WW*, 26 March 1937, p. xiv.

[16] Johnson, Lesley 1983b, 'Images of radio: The construction of the radio audience by popular radio magazines', *Melbourne Working Papers*, p. 45.

[17] *WW*, 26 December 1930, p. 6.

[18] Jones, Colin 1995, *Something in the Air: A History of Radio in Australia*, Kangaroo Press, Sydney, p. 36.

[19] Marchand, *Advertising the American Dream*, p. 108.

[20] *WW*, 5 June 1931, p. 11.

[21] Johnson, 'The intimate voice of Australian radio', p. 48.

[22] *WW*, 13 November 1936, pp. xvii–xix; 9 December 1938, p. 17.

[23] Johnson, 'The intimate voice of Australian radio', p. 48.

[24] *WW*, 27 September 1935, p. 23.

[25] *Radio Pictorial of Australia (RPA)*, 1 May 1936, p. 10; 1 July 1937, p. 15.

[26] *WW*, 27 January 1928, p. 5.

[27] Marchand, *Advertising the American Dream*, p. 93.

[28] *WW*, 10 July 1936, p. 13. See also *RPA*, 1 December 1935, pp. 15, 36; 1 November 1941, p. 37.

[29] *RPA*, 1 December 1935, pp. 15, 36.

[30] Johnson, 'Images of radio', p. 45 and Johnson, Lesley 1988, *The Unseen Voice*, Routledge, London, pp. 88, 110–11, 196, 203.

[31] Matthews, *Dance Hall and Picture Palace*, pp. 134–5. For 6ML, see *Who's Who in Broadcasting in Western Australia*, 1935–36, p. 14.

[32] Grahame, Rachel 2002, 'Eunice Minnie Stelzer', in John Ritchie and Diane Langmore (eds), *Australian Dictionary of Biography (ADB)*, Vol. 16, Melbourne University Press, Melbourne, p. 300; *WW*, 4 November 1927, p. 8; 18 September 1931, p. 8; 16 March 1934, p. 8; *Happiness*, May 1949, pp. 9–11.

[33] Eurobodalla Homes Charitable Organisation Records (EHCOR), 2GB Happiness Club Register of Members, pp. 1–2; 2GB Happiness Club (No. 2) Alphabetical Register, pp. 174–5. *WW*, 28 August 1931, p. 12; 18 September 1931, p. 8.

[34] Matthews, *Dance Hall and Picture Palace*, p. 137.

[35] *WW*, 28 August 1931, p. 12; 18 September 1931, p. 8. See also EHCOR, 2GB Happiness Club Bankstown Branch Minute Book 1932–47, pp. 13, 27, 37, 40.

[36] *WW*, 5 August 1932, p. 8; 19 August 1932, p. 29; 21 April 1933, p. 10.

[37] Johnson, *The Unseen Voice*, p. 111.

[38] Letter from Stelzer to Lady Gowrie, 4 August 1937, National Archives of Australia, ACT, A2880, 20/6/68. See also *Happiness*, May 1948, p. 42.

[39] *WW*, 20 April 1934, p. 9; Bell, Ron 1997, *The History of Radio 2GF 1933–1997*, Bell Publishing, Grafton, pp. 127–32.

[40] *Cairns Post*, 4 August 2006, p. 19.

[41] *Broadcasting Business Year Book*, 1939, p. 129.

[42] *WW*, 18 May 1934, p. 8.

[43] Towler, David J. 1984, *The First Sixty Years 1924–1984, 5DN 972*, 5DN, Adelaide, pp. 43–5.

[44] Swan Hill Regional Library, 3SH Women's Club 50th birthday article; 3SH Women's Club constitution, 3 June 1939.

[45] State Library of Tasmania Launceston Reference Library, 7LA Women's Association Records.

[46] *Report of the Joint Committee on Wireless Broadcasting*, 1942, p. 53; *Broadcasting Business Year Book*, 1939, p. 129.

[47] *BB*, 9 May 1940, pp. 16–17.

[48] Johnson, *The Unseen Voice*, p. 115; *Macquarie Network Catalogue*, n.d. (1942), pp. 22, 46, 48.

[49] The following is drawn from correspondence and scrapbooks in Arts Centre, Performing Arts Collection, Melbourne, Nancy Lee Collection. See also *Argus*, 13 September 1956, p. 14; Messenger, Dally 2002, 'Clifford Nicholls Whitta', in John Ritchie and Diane Langmore (eds), *Australian Dictionary of Biography*, Vol. 16, Melbourne University Press, Melbourne, p. 543; Lee, Nancy 1979, *Being a Chum was Fun*, Listen & Learn Productions, Melbourne.

[50] Marchand, *Advertising the American Dream*, pp. 96, 336.

11. Talking Salvation for the Silent Majority: Projecting new possibilities of modernity in the Australian cinema, 1929–1933

Brian Yecies

This chapter analyses the distinctiveness of the coming of permanent sound (the 'talkies') to the Australian cinema in the late 1920s and early 1930s. The coming of sound resulted in fundamental, but not uniform, change in all countries and in all languages. During this global transformation, substantial capital was spent on developing and adopting 'modern' technology. Hundreds of new cinemas were built; tens of thousands were wired with sound equipment—that is, two film projectors with sound attachments, amplifiers, speakers and electrical motors—and some closed in financial ruin during the Great Depression. The silent period ended and sound became projected as a symbol of progress in the metropolis and beyond. As Nowell-Smith and Ricci, and Higson and Maltby point out,[1] this incredible shift was driven principally by Western—in other words, American and European—film producer-distributors who retained a dominant influence because of their individual, and at times common, efforts to 'modernise' their production, distribution and exhibition methods.[2] The coming of sound, however, also gave rise to the development of local responses to these technological trends. It is the aim of this study to show how the 'global' transition was more local than previously thought, challenging conventional assumptions about global and local business interests in the cinema industry.

The coming of sound to the Australian cinema was pioneering and exuberant, and attitudes towards the so-called new medium were bold. The case studies of the most active of Australia's sound pioneers are presented here in order to illuminate the larger processes concerning the ambivalent embrace of modernity in Australia. Australian inventors and American electrical companies employed by the major Hollywood studios each espoused a different agenda in the promotion of their sound-projection technology.[3] I suggest that *Film Weekly* and *Everyones*—Australia's major film industry trade magazines at the time—promoted the coming of the talkies as a technological battle. They leave us with the impression that Australians were underdogs in some kind of 'talkie war', pushing an agenda of resistance against the perceived Americanisation of modernity in Australia. Opposition revolved mainly around Western Electric's plans to wire major city cinemas with American technology while preventing competition from local competitors.[4]

This confrontation spawned one of the many changes in the spheres of cinema and entertainment, which, according to Jill Julius Matthews, 'scattered the sparks of modernity into everyday life'.[5] The cost of the Western Electric system—one of the only commercially viable sound systems available in Australia in late 1928 and early 1929—remained, however, out of reach for a majority of independent, suburban and country cinemas.[6] The minimum cost of installing Western Electric sound equipment in Australia was put at between £6,000 and £7,000, which did not include the weekly fees in the company's compulsory 10-year service contracts (estimated to cost up to an additional £4,000).[7]

In response to a lack of practical and affordable choices, a wide range of Australian tinkerers, electricians, engineers, radio and wireless specialists and entrepreneurs developed their own sound systems and/or adapted acquired technology in order to bring sound to as many cinemas as possible. The geographical specificity of these producers highlights the tension between viable local and foreign technology as sound was spread across Australia's vast distances.

The US film industry in the middle to late 1920s, according to Gomery, embraced sound technology primarily for economic reasons,[8] although it could have been that the Australian film industry had little choice in the matter. Australia's largest exhibition chains committed ambitious investments to convert to sound with US companies in order to screen Hollywood films. That is to say, the distribution side of the Australian film industry was dominated largely by American interests, making the Australian market one of the largest sources of foreign film revenues for major Hollywood studios and their producer-distributors. When the local distribution exchanges of Fox, First National-Warner Bros., FBO-RKO, MGM and Paramount—all controlling members of the Motion Picture Distributors Association (MPDA) of Australia—began regularly adding sound films (that is, all-talkie, part-talkie and films with sound effects, asynchronous and/or synchronous music) to their catalogues in 1929, it seemed as though the conversion to sound was inevitable, at least initially for capital cities.[9] At the same time, Western Electric and the powerful members of the MPDA of Australia appeared to thwart competition at all costs. Therefore, Australian sound companies and inventors developed strategic methods and alliances in order to meet international and domestic competition head-on.

Australia's 'talkie war'

The concept of marrying sound and pictures in Australia had been around since the start of cinema. The coming of sound was not a new phenomenon, as many popular accounts would have it. Collins proposes that the Australian cinema was never silent—'some form of sound accompaniment dates back to the origins of moving pictures'.[10] Live musical performances enhanced the cinematic experience, creating extra value for spectators, though mechanically reproduced sound eventually became cheaper than an orchestra, especially for the larger city cinemas. Until the late 1920s, sound technology developed outside the Hollywood studio system was promoted as a temporary and novel attraction. Yet, the Australian cinema experienced a new beginning in late 1924 when entrepreneurial picture showmen imported and endorsed American De Forest Phonofilms' sound-on-film projection equipment. Until this time, projecting silent films with live accompanying music or recorded music and dialogue on large phonograph discs was the most common method of presenting a sound film.

De Forest Phonofilms Australia Ltd (Phonofilms Australia), a sound company operating independently of the Hollywood majors, was formed in August 1925 (first through American Phonofilms and then through British Phonofilms).[11] As Figure 1 illustrates, the company promoted its novel system as the ultimate sign of modernity: 'One of the most amazing achievements of the age'—aiming to build a local franchise and to have its technology adopted. As seen in other trade advertisements, Phonofilms Australia wanted exhibitors and industry people to know that it was leading the 'dawn of a new era in the industry'.[12]

American distributors did not appreciate alternative and ambitious efforts to modernise Australia with sound technology outside the Hollywood studio system. As I have written in more detail elsewhere, it was Phonofilms Australia to which the Australian press attributed the instigation of the talkie war.[13] Headlines in the trade magazines screamed: 'War of "Talkies" Looms as Phonofilms Issue Warning.'[14] The general managers of Phonofilms Australia had threatened legal action against anyone using a sound-on-film process that infringed Phonofilms' rights in Australia.[15] Nevertheless, a host of local competitors soon attempted to challenge Phonofilms. Suffice to say that the De Forest Phonofilms Company advanced the experimentation of film sound technology and the acceptance of that technology in Australia by setting the stage for modernity and the further adoption of sound innovations.

Building on this Phonofilms Australia narrative is the arrival in Australia in November 1928 of sales and service engineers from Western Electric-ERPI, who came to wire simultaneously the Hoyts Theatres' Sydney Regent (with Fox Movietone brand sound-on-film equipment) and the Union Theatres' Sydney Lyceum (with Vitaphone brand sound-on-disc equipment). The public had the greatest opportunity to experience (to see and hear) urban modernity at Union Theatres' and Hoyts Theatres' flagship cinemas because this was where the industry's changes first occurred.

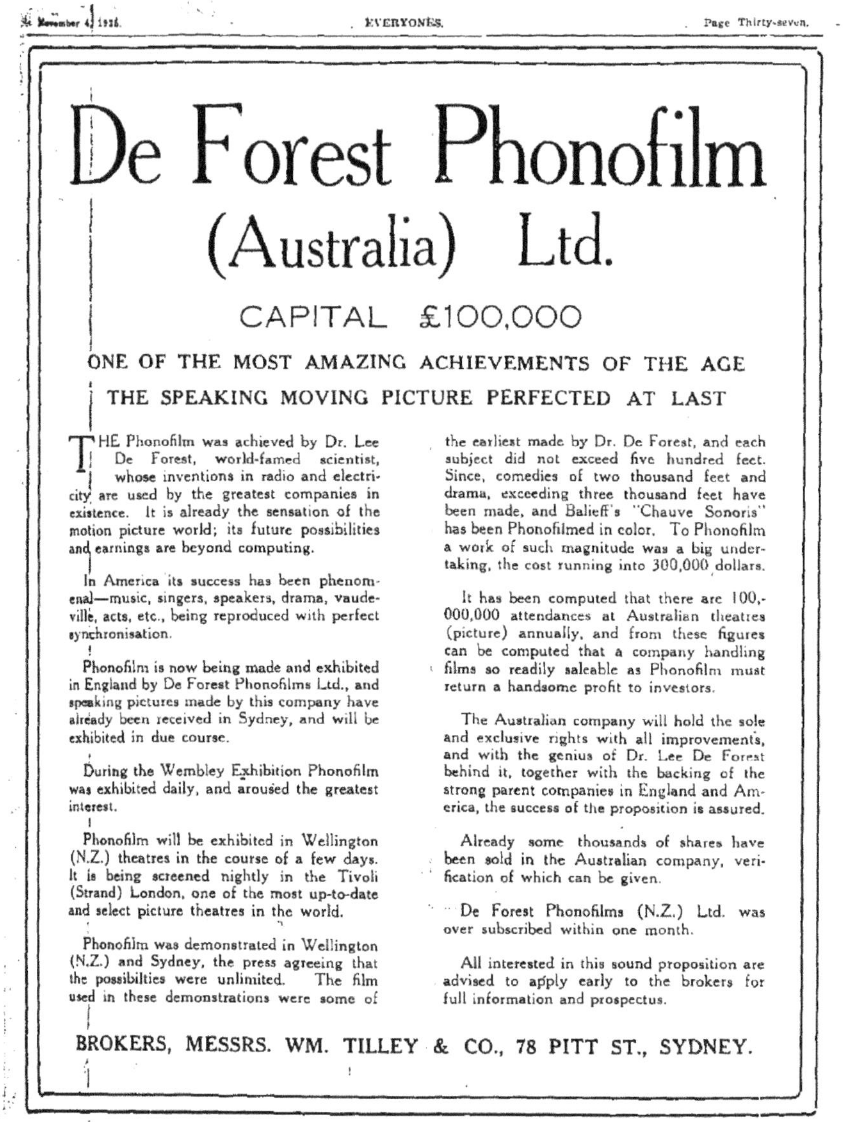

De Forest Phonofilm (Australia) Ltd.

CAPITAL £100,000

ONE OF THE MOST AMAZING ACHIEVEMENTS OF THE AGE
THE SPEAKING MOVING PICTURE PERFECTED AT LAST

THE Phonofilm was achieved by Dr. Lee De Forest, world-famed scientist, whose inventions in radio and electricity are used by the greatest companies in existence. It is already the sensation of the motion picture world; its future possibilities and earnings are beyond computing.

In America its success has been phenomenal—music, singers, speakers, drama, vaudeville, acts, etc., being reproduced with perfect synchronisation.

Phonofilm is now being made and exhibited in England by De Forest Phonofilms Ltd., and speaking pictures made by this company have already been received in Sydney, and will be exhibited in due course.

During the Wembley Exhibition Phonofilm was exhibited daily, and aroused the greatest interest.

Phonofilm will be exhibited in Wellington (N.Z.) theatres in the course of a few days. It is being screened nightly in the Tivoli (Strand) London, one of the most up-to-date and select picture theatres in the world.

Phonofilm was demonstrated in Wellington (N.Z.) and Sydney, the press agreeing that the possibilties were unlimited. The film used in these demonstrations were some of the earliest made by Dr. De Forest, and each subject did not exceed five hundred feet. Since, comedies of two thousand feet and drama, exceeding three thousand feet have been made, and Balieff's "Chauve Sonoris" has been Phonofilmed in color. To Phonofilm a work of such magnitude was a big undertaking, the cost running into 300,000 dollars.

It has been computed that there are 100,000,000 attendances at Australian theatres (picture) annually, and from these figures can be computed that a company handling films so readily saleable as Phonofilm must return a handsome profit to investors.

The Australian company will hold the sole and exclusive rights with all improvements, and with the genius of Dr. Lee De Forest behind it, together with the backing of the strong parent companies in England and America, the success of the proposition is assured.

Already some thousands of shares have been sold in the Australian company, verification of which can be given.

De Forest Phonofilms (N.Z.) Ltd. was over subscribed within one month.

All interested in this sound proposition are advised to apply early to the brokers for full information and prospectus.

BROKERS, MESSRS. WM. TILLEY & CO., 78 PITT ST., SYDNEY.

Figure 1

Source: Advertisement, De Forest Phonofilms Australia, *Everyones*, 4 November 1925, p. 37. Reprinted by permission of the National Library of Australia.

In late December 1928, Warner Bros.' *The Jazz Singer* premiered at Union Theatres' Lyceum and Fox's *The Red Dance* premiered at Hoyts' Regent, sending shock waves through Sydney and the whole of the film industry. According to reports in *Everyones* and *Film Weekly*, the age of sound arrived in a 'big way'.[16] In one week, the Lyceum and Regent combined took between £8,000 and £8,500—breaking all records and unexpectedly selling out all shows.[17]

According to the trade papers, talkies had rejuvenated public interest in motion pictures and rescued the film industry from its pre-Depression era slump.[18]

Suburban exhibitors immediately began wondering how soon they could convert to sound and modernise their programs. Acting as the 'voice' of the Australian film industry, however, Stuart Doyle, the managing director and primary spokesman of Union Theatres Ltd—and powerful manager of Australasian Films Ltd—pragmatically cautioned exhibitors about the future of talkies in the suburbs:

> No theatre can show talking pictures profitably unless it can take from £1,200 to £2,000 a week; and unless some method is devised to cut down the cost of equipment, reproducing records and particularly the terrific cost of maintenance, talkies are a commercial impossibility for suburban and small town shows.[19]

So suburban and country exhibitors waited, but not for long.

In mid-1929, the talkie war campaign resumed in the trade papers when the American RCA Photophone and a few Australian sound systems—such as Raycophone, Australtone, Markophone and Auditone—entered the field and began competing with Western Electric for installation contracts and equipment leases and sales.[20] The talkie war, which had started in 1927 with Phonofilms Australia, had not been forgotten.

In Australian film history—and probably in that of many other countries—there remains a dearth of information surrounding the coming of sound. Previous studies provide brief summaries of activities between late 1928 and mid-1929, while giving scant attention to the details surrounding entrepreneurial efforts before and after this time.[21] Shirley and Adams stress that Australian exhibitors converted to sound in order to survive as 'new pioneers' attempted 'to break the [US] monopoly' by developing sound recording and projecting equipment of their own.[22] Tulloch presents the idea of a hierarchical reaction to the apparent American dominance of sound technology, which was 'itself a hierarchy established in, and exported from, the USA'.[23] Walsh's more recent work stands apart by offering an examination of the opposing relationships between distributors and exhibitors, entertainment moguls, manufacturers and users of technology, as well as industry and government.[24] He indicates that there was more than a simple American monopoly during the transition; Hollywood did not wield 'ruthless power' over Australia. Walsh coins the phrase 'frenzy of the audible', which captures the spirit of modernity in Australia as some kind of sound technology became part of the exhibition experience before, during and after the 1920s.[25] With these studies in mind, it is unmistakable that Hollywood's efforts in the export (transfer) of US technology and infrastructure for sound came to a head in 1929—'a year of sound and fury', which proved that 'silent cinema was doomed'.[26]

Preventing extinction of the Australian enterprise

Primary research coalesced in Figure 2 shows at least 67 Australian sound systems that challenged the apparent American dominance. The geographical rise and development of these alternative sound systems is noteworthy because it suggests that sound technology was spread by a multitude of sound companies, engineers, tinkerers and theatre and cinema entrepreneurs. Despite the portrayal in the trade press of Hollywood film distributors as military victors in a so-called war to wire Australia with American technology, these Australians invigorated—that is, mediated—modernity throughout a vibrant period of sound experimentation. Not all initiatives were successful, but clearly the demand for sound was heard all over Australia.

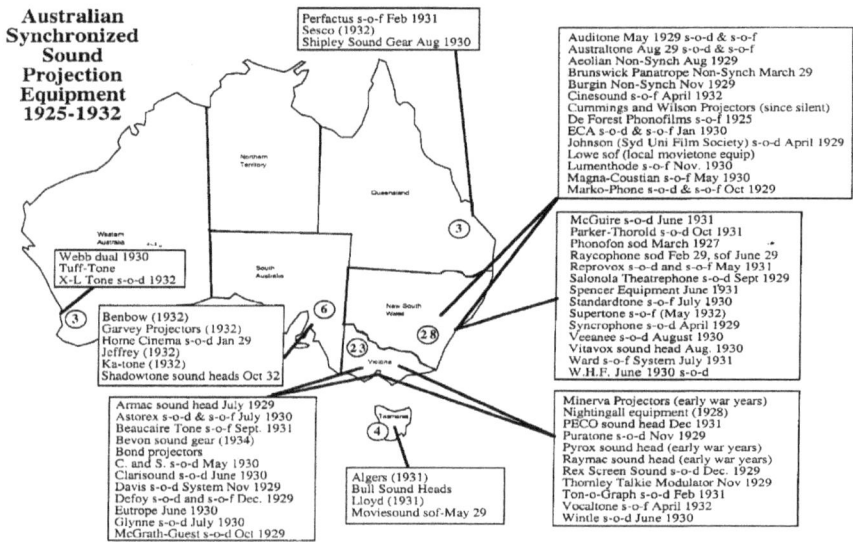

Australian Synchronized Sound Projection Equipment 1925-1932

Perfactus s-o-f Feb 1931
Sesco (1932)
Shipley Sound Gear Aug 1930

Auditone May 1929 s-o-d & s-o-f
Australtone Aug 29 s-o-d & s-o-f
Aeolian Non-Synch Aug 1929
Brunswick Panatrope Non-Synch March 29
Burgin Non-Synch Nov 1929
Cinesound s-o-f April 1932
Cummings and Wilson Projectors (since silent)
De Forest Phonofilms s-o-f 1925
ECA s-o-d & s-o-f Jan 1930
Johnson (Syd Uni Film Society) s-o-d April 1929
Lowe sof (local movietone equip)
Lumenthode s-o-f Nov. 1930
Magna-Coustian s-o-f May 1930
Marko-Phone s-o-d & s-o-f Oct 1929

McGuire s-o-d June 1931
Parker-Thorold s-o-d Oct 1931
Phonofon sod March 1927
Raycophone sod Feb 29, sof June 29
Reprovox s-o-d and s-o-f May 1931
Salonola Theatrephone s-o-d Sept 1929
Spencer Equipment June 1931
Standardtone s-o-f July 1930
Supertone s-o-f (May 1932)
Syncrophone s-o-d April 1929
Veeanee s-o-d August 1930
Vitavox sound head Aug. 1930
Ward s-o-f System July 1931
W.H.F. June 1930 s-o-d

Webb dual 1930
Tuff-Tone
X-L Tone s-o-d 1932

Benbow (1932)
Garvey Projectors (1932)
Home Cinema s-o-d Jan 29
Jeffrey (1932)
Ka-tone (1932)
Shadowtone sound heads Oct 32

Minerva Projectors (early war years)
Nightingall equipment (1928)
PECO sound head Dec 1931
Puratone s-o-d Nov 1929
Pyrox sound head (early war years)
Raymac sound head (early war years)
Rex Screen Sound s-o-d Dec. 1929
Thornley Talkie Modulator Nov 1929
Ton-o-Graph s-o-d Feb 1931
Vocaltone s-o-f April 1932
Wintle s-o-d June 1930

Armac sound head July 1929
Astorex s-o-d & s-o-f July 1930
Beaucaire Tone s-o-f Sept. 1931
Bevon sound gear (1934)
Bond projectors
C. and S. s-o-d May 1930
Clarisound s-o-d June 1930
Davis s-o-d System Nov 1929
Defoy s-o-d and s-o-f Dec. 1929
Eutrope June 1930
Glynne s-o-d July 1930
McGrath-Guest s-o-d Oct 1929

Algers (1931)
Bull Sound Heads
Lloyd (1931)
Moviesound sof-May 29

Figure 2

Source: Derived from articles and advertisements found in *Film Weekly* and *Everyones* between 1925 and 1932.

The Raycophone case exemplifies how at least one eminent Australian system achieved 'national' status through the strategic reactions to the importation of American sound technology. The Australian-made Raycophone ('Rayco' from its inventor, Raymond Cottam Allsop's, name and 'phone' from its inclusion of disc technology) outperformed and outlived all other Australian-made sound systems. Allsop's survival strategy began with price—promising a dual-disc and sound-on-film system for a 'revolutionary' cost of £1,500.[27] Raycophone was given several stealthy trade demonstrations in early February 1929, and was first demonstrated publicly on 10 June 1929.[28] The company's take-over in April 1930 by Harringtons Ltd—an Australian photographic (and later film) equipment supplier and distributor since 1899 and a radio supplier since the inception of the industry in 1888—further ensured its survival by increasing Raycophone Ltd's manufacturing and engineering talent, enlarging its sales and distribution potential and infusing financial resources into Allsop's regional sound company.[29] Its success was due to the company's political and film industry ties, which led ultimately to its 24 per cent market penetration. As of June 1937, 345 Raycophone sound-projection systems were installed and serviced in Australia,[30] demonstrating that the coming of sound to the Australian cinema was facilitated directly by employing Australian technology and Australian engineering as much as it was by American technology.[31]

Figure 3

Source: Advertisement, Raycophone, *Everyones*, 12 June 1929, pp. 26–7. Reprinted by permission of the National Library of Australia.

After the launch of Western Electric's sound-projection system in Australia in November 1928, Allsop began developing a mechanism for projecting sound films in an auditorium, using amplifiers and loudspeakers.[32] Yet, as a radio pioneer and prominent Sydney electrician, he saw the likelihood of the advent of motion picture sound in Australia well before Western Electric-ERPI representatives set sail for the region. Allsop was an electrician in charge of designing and building transmitters for 2BL, which was a New South Wales radio station owned by Broadcasters-Sydney Ltd—one of the earliest radio stations to go to air in Australia, in November 1923.[33] According to his own testimony in an ABC radio interview in 1970, Allsop had begun developing sound-film technology in the early 1920s in experiments that were extensions of his radio knowledge. As early as 1921, Allsop claimed to have successfully synchronised a wax cylinder recording with a motion picture film, and there is no reason to doubt him. Allsop had also produced four short synchronised vaudeville films—shot in the old 2BL radio studio overlooking Market Street in Sydney—in order to prove the capabilities of his disc system.[34] Allsop ceased his sound-film experiments after 1921, however, primarily because of limitations in audio amplification.

When Allsop resumed his experiments seven years later in June 1928, there seemed to be one key reason for him getting back into sound: Warner Bros.' success with *The Jazz Singer* in October 1927. By this time, the development of efficient loudspeaking mechanisms and improvements in audio amplifiers significantly enhanced the profitability of the talkies in the United States. In addition, there were 28 American talkie features in Australia in early January 1929 and only three cinemas wired for sound.[35] Furthermore, in early February 1929, headlines declared, 'Talkie field now open for all quality sound devices', which was in response to Western Electric's (interchangeability) decision in the United States to allow competing systems—which Western Electric put through a rigorous approval process—to project Hollywood films.[36] In this way, Allsop and many other Australians had more than ample motivation to perfect an Australian-made sound-projection system.

Raycophone's supporters and company sponsors were a network of influential political contacts and well-known theatre and cinema entrepreneurs—basically friends in high places whom Allsop exploited for obvious reasons. The list includes the Australian entertainment moguls J. C. Williamson Ltd, Keith Murdoch (media tycoon Rupert Murdoch's father), the Tait brothers, Farmer & Co. Ltd and the NSW Broadcasting Company (NSWBC), which represented an influential association of Australian media organisations. The Williamson–Tait group was connected with Allan and Company (music distributors) and Buckley and Nunn Ltd (radio and amplification equipment retailers).[37] Murdoch brought invaluable support to the Raycophone organisation from his *Herald and Weekly Times*, publishers of the *Sydney Morning Herald*, the *Sun* and *Telegraph Pictorial*.[38] This web of interrelated connections extended throughout the Australian motion picture and entertainment industries, linking Raycophone to Western Electric, Union Theatres and Movietone, via Hoyts Theatres.

Australian systems competed with one another in addition to international representatives. For example, in mid-1929, the Australtone sound company claimed that it planned to produce a series of short sound films in order to supply its installation clients with continuous entertainment.[39] Australtone began as a disc system, and was first demonstrated publicly to industry representatives and newspapermen in the Queen's Theatre in Crow's Nest in Sydney on 31 July 1929. This was about seven weeks after industry magazines claimed that the talkie war had been declared in Sydney, with Raycophone in the field. William J. Tighe was the young ambitious engineer credited with inventing Australtone and synchronising its pictures, which the press coverage indicated was an insurmountable challenge in the face of Western Electric's market strength and overall wiring plans for Australia.[40] Tighe's system was under great pressure to prove that it could handle disc films with the same level of quality as that of Western Electric's Vitaphone, as well as Raycophone. A series of 12 talking films referred to as 'local favourites', made by the 'world-class' cinematographer Arthur Higgins, were used to debut the disc system along with pre-recorded music. Higgins had filmed the pictures for Australtone while Tighe innovated the system during the first half of 1929. His reputation was used to help launch Australtone as a world-class producer of sound films.[41]

While Allsop took about four months of canny planning and publicity as well as sound engineering to attract the financial backing and support needed to launch Raycophone as a competitor to Western Electric, Tighe and his relative Lewis C. Tighe, an advertising specialist, appeared to do the same in four weeks with Australtone. Australtone quickly rose in the shadow of Raycophone, but it then experienced a swifter fall.

Unlike Raycophone, Australtone did not have friends in high places. Apart from minor quality concerns, a lack of influential connections inside and outside the film industry hindered its potential to offer viable long-term solutions to smaller suburban and country exhibitors. Although the Tighes promoted Australtone as the 'salvation of suburban men', American film distributors refused to supply the system with their sound films and boycotted it in late 1929.[42] This dispute manifested itself as exhibitors protested the huge expenses involved in wiring their cinemas with American sound equipment (see Figure 4). As a result of the boycott, Australtone began looking to South Australia and Western Australia, where a delay in the coming of sound provided additional opportunities for installation contracts—particularly for disc equipment. At the end of 1930, South Australia reportedly had the highest percentage of disc installations—30 out of a total of 54 (56 per cent)—while Western Australia had 26 disc-only and 14 dual projectors, reflecting the nature of experimentation in those states as much as the distribution speed of disc equipment.[43] Although Australtone disappeared by 1933, its story is symbolic of the complexities of the diffusion of the coming of sound to the Australian cinema and the cutthroat competition that existed between Australian-made and international equipment.

Figure 4

Source: Advertisement, Australtone–Australian Synchronised Sound Pictures, *Everyones*, 30 October 1929, pp. 18–19. Reprinted by permission of the National Library of Australia.

While Raycophone and Australtone began their installation campaigns in NSW, the Australian-made and 'all-Australian talkie gear' Markophone dual-disc and sound-on-film system was publicised heavily to Victoria's regional exhibitors in 1929 and 1930. By the beginning of August 1929, Western Electric-ERPI had installed equipment in only 13 of Victoria's estimated 107 cinemas[44] —hence the market seemed wide open for competition. Promoters endeavoured—as with nearly all other sound systems—to publicise and demonstrate Markophone (as seen in Figure 5) in advance of its availability, creating awareness and interest in regional Victoria, which flowed into NSW.

Adaptability to existing silent projection equipment was one of Markophone's key features. According to company records, Markophone was a 'sound head' and 'synchronised disc attachment plan' created by Hoyts Theatres' cinema executives Leon Samuel Snider, Edward Lewis Betts and Alexander Henry Noad—investors in the system. Snider, Betts, Noad and Betts' brothers (Edward, George and Frederick) were also large Hoyts shareholders. The names of Snider, a well-known exhibition capitalist, and the Betts family, well-known NSW exhibitors since before 1920, added a significant contribution to Markophone Ltd because of their ties to Hoyts.[45] This support from Hoyts Theatres Ltd executives—including engineering support from Hoyts technicians—helped disseminate Markophone until the mid-1930s, meaning it survived significantly longer than most other Australian sound systems.[46] The company eventually folded in the mid-1930s, when most cinemas had been wired for sound. Apparently, the company had fulfilled its central aim of servicing the transition to sound.

THE MOST COMPACT AND MECHANICALLY PERFECT AUSTRALIAN EQUIPMENT THAT CAN COMPETE WITH ANY IMPORTED SYSTEM AND THE ONLY AUSTRALIAN TALKIE EQUIPMENT THAT EMBODIES STANDARD ENGINEERING PRINCIPLES.

MARKOPHONE

SOUND ON FILM
SOUND ON DISC   ÷ REPRODUCING SYSTEM   SOUND ON FILM
SOUND ON DISC

UNIVERSAL BASE ADAPTED TO SIMPLEX—C. & W.—POWERS—KALEE—HAHN-GOERZ, ETC. PROJECTORS.

10¡ Points for¡Markophone

(1) Scientifically designed—no breakdowns.
(2) Latest filteration system—no ground noise.
(3) Patent tone arm pick-ups—non-slip tracker.
(4) Variable speed motors—for silent pictures.
(5) Dual monitor control—eliminates extra staff.
(6) Exponential horns—no baffle board screechers.
(7) Engineering perfection—highest qualifications.
(8) Two turn-table non-synchronous equipment.
(9) PHILIPS AMPLIFICATION SYSTEM (renowned throughout the universe).
(10) Spares with every equipment.

MARKOPHONE

THEATRE ROYAL. WEST RYDE, N.S.W.; PARAMOUNT THEATRE. MORTDALE, N.S.W.; PALACE THEATRE. RYDE, N.S.W.; NATIONAL THEATRE. RICHMOND, VIC.

LOCO THEATRE. NORTH MELBOURNE, VIC.; CITY COUNCIL THEATRE. WILLIAMSTOWN, VIC.; ST. GEORGE THEATRE, YARRAVILLE, VIC.

AND THE SAVOY (LATE ADYAR HALL) SYDNEY

OPENS THIS MONTH FOR CONTINUOUS TALKIES WITH MARKOPHONE.
MORE THAN 20 MARKOPHONE EQUIPMENTS WILL BE INSTALLED BY END OF JULY.

PRICE £1450   PRICE £1450

TERMS FROM £100 DEPOSIT.

BALANCE OVER TWO OR THREE YEARS.
WE SELL—NOT LEASE—IT BECOMES AN ASSET OF THE THEATRE.
£50 WILL COVER WIRING. SCREEN FRAME, INSTALLATION. ETC.
IF YOU HAVE A NON-SYNCHRONOUS SET WE WILL MAKE A REASONABLE ALLOWANCE.

NO FEE SERVICE   SERVICE NO FEE

Sound experts agree that an operator after two weeks tuition is efficient in adjustment of minor defects of not only Markophone but any talkie equipment—Why a set service fee?
Our service staff at all hours available——periodical inspections.
MARKOPHONE IS FOOL PROOF—Pay for Service You Require and Receive.

MARKOPHONE LTD., 521 George St., SYDNEY.   London Stores Building, MELBOURNE.

Figure 5

Source: Advertisement, Markophone, *Everyones*, 4 June 1930, p. 49. Reprinted by permission of the National Library of Australia.

Finally, between 1929 and 1931, Phonofilms Australia inspired another Australian sound system: the Auditone, and its many incarnations.[47] Charles Ward, a radio engineer from New Zealand who gained valuable training and experience working on the De Forest Phonofilms system in 1925, developed the Auditone in conjunction with Stanley William Hawkins, the entrepreneur who managed Phonofilms Australia in 1927.[48] Auditone was demonstrated, promoted and recast several times as a new system (with names such as First British, Ward and 'Lumenthode-Beam Projection and Sound Apparatus'), yet failed as a result of tough competitive tactics by Western Electric. The alternative equipment showed promise by gaining a number of installation contracts, but it lost its major sponsor towards the end of 1929 when Broadway Theatres, a regional chain of cinemas, cancelled its Auditone contracts.[49] As a result of these losses, Auditone Ltd lost a significant window of opportunity before the onset of the Great Depression, which surely played a role in its liquidation proceedings in 1931.[50]

Despite what Australian histories tell us, inventors across the nation moved local manufacturing forward by pooling talents, wireless and phonography expertise, entrepreneurial visions and scientific research, forming an active national sound-film industry. The existence of many of the Australian sound companies and alternative systems pictured in the map in Figure 2 were based first on the level of investment capital used to finance new wiring and equipment manufacturing ventures, and second on the cost of the equipment. Outdated (or unwanted) technology could be sold off to smaller, suburban and rural cinemas as larger city cinemas upgraded to newer models, which became competitive advantages against other systems and exhibitors. This is a critical point because the cinemas that could not afford the latest sound technology used disc systems for much longer than most city venues in order to save money. For all of these reasons, the transition to sound was an uneven process, which took place in a seemingly unorganised way.

After 1929, Western Electric and RCA continued to wire a majority of the larger metropolitan and suburban cinemas, replacing many of the Australian systems identified in the map in Figure 2 that did not meet 'American quality standards'. These systems were eventually thrown on what the trade press frequently called the scrap heap. In short, by 1937, all of Australia's cinemas had been wired,[51] and all of the Australian-made systems had perished except for the Raycophone, which lasted well into the 1950s.

'Quality' of local, regional and national ingenuity

Ostensibly, the coming of sound seems to have facilitated Hollywood's dominance of English-speaking countries, raising sharp issues about the promotion and protection of foreign versus local interests. It is true that between 1929 and 1932 Australians utilised a combination of new and improved disc and sound-on-film systems as a means of circumventing US patent monopolies. Rather than group all Australian sound-reproducing equipment into one category, however—as the film industry trade papers did with the terms 'local' and 'locally made'—it might be more productive to think about this history more diversely in terms of local, regional and national specificity.

A range of Australian-made sound projectors appeared as near copies of existing systems with little chance of a successful patent application, or an imported system promoted as a local product. Projection technology, then, was customised and modified (adapted) on a local level by local engineers after arriving at the installation location. Local systems can be defined as backyard innovations and one-off adaptations assembled for use in schools, country town halls, churches and travelling shows. Country-based systems were made from scratch—primarily in the disc format—representing the cheapest ways to exhibit sound films. Some were promoted by smaller, often fly-by-night organisations. All were alternatives to the dominant Western Electric and RCA equipment and to the better-known Australian equipment—Raycophone, Australtone, Markophone and Auditone—combining new and used phonograph and projector parts, motors, gears and belts (rubber and leather), connecting a gramophone player or another type of disc turntable with a silent projector. Designers were usually the operator and sole service/maintenance engineer, since only they knew how the modifications worked. Such people projected for themselves, friends or for managers of smaller venues. One local device called the Dowling Apparatus, which 'causes the gramophone and the film to synchronise exactly', cost as little as £5.[52] Sadly, little is known about its Mr Dowling and many other pragmatic and opportunistic tinkerers like him.

Creators of so-called locally made or alternative sound-projection systems were under enormous pressure to promote any and all competitive advantages regarding their particular equipment. Claims of local origins were often prominent in public relations articles and advertisements: Wintle, 'Australian Made, 100% British';[53] Astorex, 'All Australian Talkie Equipment';[54] Beaucaire, 'Made and Patented in Australia';[55] and Australtone, 'All Australian Talkie Equipment'.[56] Paradoxically, the above claims attempted to reaffirm the idea of a national sound industry by the inventors or innovators themselves, yet few could secure successful patent applications, which represented a primary avenue for the recognition and protection of national inventions. These clever local concoctions formed the lower levels of a hierarchy of sound equipment in Australia.

'Regional' Australian-made equipment, in comparison, cost more than £5, but was mostly still affordable. Slick trade advertisements promoted installations of disc systems at a cost of £225–300 for a single apparatus, often omitting prices for a complete system (which was two projectors and two turntables for reel change-overs) and making prices appear more attractive. As seen in Figure 6,

Bert Cross and Arthur Smith's Cinesound sound-on-film system was advertised 'from £225'.[57] Exhibitors had to double these prices in order to get a complete system.

Figure 6

Source: Advertisement, General Theatre Supplies Ltd—Cinesound, *Film Weekly*, 25 April 1932, p. 19. Reprinted by permission of the National Library of Australia.

Regional inventions such as Cinesound were constructed by industrial systems, produced in greater numbers and distributed to several exhibitors and/or multiple installations in cinemas throughout one particular state. They were more durable than local systems because they were often cast in metal. Unlike local sound-on-disc systems, a majority of regional systems offered sound-on-film as well. Not all of a brand's units looked the same, however, because design modifications were necessary due to the diversity of the pre-existing silent equipment to which they were adapted. These systems, which often employed a team of service/maintenance engineers to handle large distribution contracts, made up the middle hierarchy of sound equipment in Australia.

On the whole, Australian sound companies shifted positions in the hierarchy. For example, Auditone Ltd was a regional system until it opened its second sales branch in Brisbane in June 1930, at which point Auditone became a 'national' system.[58] National systems had an established distribution network in place or soon developed one in order to promote and accommodate interstate sales. In addition, national companies were supported by substantially greater finance capital than regional companies. As indicated earlier, Raycophone became a national system after 1930, when established national photographic and radio supply company Harringtons acquired Allsop's promising regional venture and transformed it into a thriving national institution. Showmen and exhibitors could now inquire about Raycophone equipment and contract information at any of the three locations operated by Raycophone Ltd in Sydney (assembly works, factory or city offices) or at one of Harringtons' many offices in Sydney, Newcastle, Katoomba, Melbourne, Brisbane, Adelaide, Perth and Hobart, as well as in Auckland and Wellington in New Zealand.[59]

As we have seen, Australian sound companies were many and varied. Entrepreneurial individuals, radio amateurs, engineers and electrical enthusiasts and well-organised companies promoted sound-on-disc and sound-on-film projection systems because no one seemed to know which format would become standard. At the same time, US film companies continued to distribute silent films until late 1932 and early 1933 in order to profit from the cinemas still waiting to convert to sound, elongating the industry's complete transformation. The coming of sound to Australia took at least a decade to achieve, and new developments and improvements are being made continually today.

Many of the models pictured and discussed in trade advertisements were created for demonstration purposes in order to judge the system's potential to generate profits for the exhibitor and sound company. For example, the Wintle disc system from Melbourne (as seen in Figure 7), which was known as both Australian-made and '100 per cent British', was promoted as a new local product, but never came to fruition because of a lack of exhibitor interest.

The advertisement below illustrates the confusion and the disorder exhibitors faced during the coming of sound. One trade report of bad sound reproduction could kill any chance of an invention's acceptance, let alone success. The designs and development of many projectors reached only a prototype stage where people could view it because they could not be or were not mass-produced. Yet, installing (and purchasing or leasing) a local, regional or national Australian projection system surely provided a kind of salvation for the suburban exhibitor because it circumvented issues of foreign patent ownership and control, high import duties, royalty fees and Western Electric's mandatory 10-year service plan. Rather than a 'battle', however, there was a notable frenzy of invention, innovation and adaptation during the transition as Australians attempted to project sound films with local technology.

Promotional campaigns for Raycophone, Australtone, Markophone and Auditone emphasised price, ownership and customer testimonials in localised contexts. While trying to survive, however, nearly all systems around the globe either succeeded or failed by living up to or failing to live up to Western Electric's insistence on a particular kind of quality—expressed in the advertising discourse and trade press campaigns run by Western Electric (and RCA, to a lesser degree). Western Electric's arrival, business strategies and intense promotional efforts can be read as overt attempts to push its quality standards at every opportunity in order to belittle competitors. As seen in Figure 8, Western Electric wanted Australians to believe that it made the 'dream' of projecting and amplifying the talkies a modern reality.

Australian systems had little choice but to imitate the capabilities of Western Electric equipment in order to project all types of Hollywood sound films. This strategy was known commonly as the 'interchangeability' factor, and was really about a system's perceived performance and service in a global context. Demanding quality from an alternative system was Western Electric's primary control device. Each sound company mentioned above had to be able to demonstrate—through trade screenings or through gaining Western Electric's approval and/or through a distributor's continuing contracts for films—that its equipment provided exhibitors with sound that was good enough to keep the patrons coming.

It seems that sound companies continually perfected their systems in this way throughout the transition to sound. The American RCA Photophone succeeded because, after July 1928, it and Western Electric had agreed to make their systems fully compatible. As a result, when RCA arrived in Australia in May 1929, the quality of its system had been accepted and endorsed by Western Electric for nearly a year. As seen in Figure 9, the RCA company promoted heavily its patented technology and aggressive service strategies.

Figure 7

Source: Advertisement, Wintle Sound System, *Everyones*, 4 June 1930, p. 43. Reprinted by permission of the National Library of Australia.

Figure 8

Source: Advertisement, Western Electric, *Film Weekly*, 1 August 1929, p. 13. Reprinted by permission of the National Library of Australia.

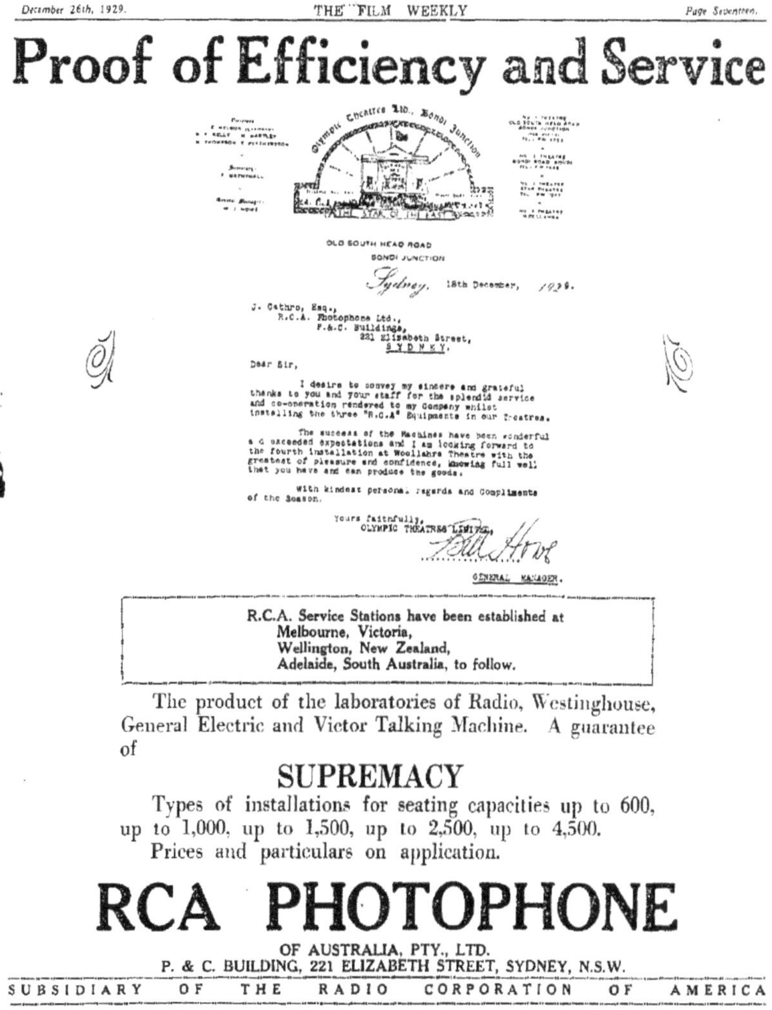

Figure 9

Source: Advertisement, RCA Photophone, *Film Weekly*, 26 December 1929, p. 17. Reprinted by permission of the National Library of Australia.

As we have seen, other systems, such as Australtone, were not as fortunate because they did not meet Western Electric's standards. Hence, a system's versatility (its ability to handle films recorded in the Movietone, Vitaphone and Photophone format) undoubtedly played a role in its success.[60]

Conclusion: Australia's experience of modernity in sound

The research in this chapter has attempted to show how the entrepreneurial ambitions of Australian sound companies and other individuals were part of

Australia's experience of modernity. The coming of sound was influenced and shaped by a diverse group of Australians who at least pretended to be modern and/or understood the line of modernity brought by the prominent distribution of thousands of American (and European) sound films. Their individual and at times shared experiments with film sound represent a kind of fluid spirit that stirred in hegemonic responses to the demands of Hollywood. That is, Australians did whatever they could to innovate and adapt sound technology as well as to facilitate the exhibition of Australian sound films[61] —all while trying to assess and manage the trade-offs between opportunity and risk. They were also caught between the promise of global modernity, collusion and the threat to national identity.

We can read their contributions as a complex combination of patriotism and conscientiousness, on the one hand, and a pragmatic and opportunistic spirit, on the other hand, which in due course enabled these representative Australians to join in the trajectory towards modernity. Whether they realised it or not, they were all agents of modernity, mediating complex social, cultural and technological changes in Australia, as were other pioneers delving into sound in other parts of the 'modern' world. Ultimately, most Australian systems and sound companies failed, but their reactions to the rhetoric of modernisation—or Americanisation—of the Australian film industry were significant in Australia's overall cultural and technological history. This is what makes the case of the coming of sound to Australia distinctive.

Evidence in the trade papers and archives suggests that the expression of Australia's modern consciousness rose partly in resistance to the 'American twang' and its potentially negative impact on Australian society. American twang was a popular expression found in major Melbourne newspapers while the talkie technology battle was taking place. During the mid-1920s and early 1930s, headlines such as 'American Films—Evil Effect on Young People',[62] 'Bad English in "Talkies"—Rigid Censorship Advised',[63] 'American "English"—Impure Accent Feared',[64] '"American Twang"—Talking Pictures Criticised',[65] 'Menace to Children'[66] and 'The Film Till Now—A Story of Americanization' riddled the popular press.[67] I interpret these articles and anonymous editorials, which were supported with distribution and exhibition statistics, as partly the result of anti-American attitudes linked to the increased use of subtitles and intertitles in silent films.

Educators and empire loyalists saw Hollywood films and American culture in general as lowering moral standards in Australia. At the same time, they saw the adoption and diffusion of talkie technology as an opportunity to promote British culture and 'pure' English.[68] Between 1929 and 1933, anxieties about potentially being Americanised surfaced in annual education conferences and teachers' union meetings. Parliamentary debates also included heated discussions on the topic. These anti-American feelings were real; however, the Federal Government and the national film censors could do little to stop the barrage of Hollywood films. Essentially, this was one of the lines of logic used to advocate for a stronger film industry throughout the Commonwealth.

In 1926, 83 per cent of films imported to Australia were from the United States, while only 7 per cent were from Britain. Hollywood's 'corrosive effect' led eventually to a Royal Commission in May 1927, which heard testimonies of thousands of witnesses from the film industry and the public, who pleaded with the Ministry of Trade and Customs to restrict the importation of US films. Increasing their tariffs, which potentially ensured fair competition, and lowering tariffs on British films were common requests. The Royal Commission lasted a year and concluded the obvious: there was an American stronghold on the film market in Australia. Little could be done, however, to reverse this trend—even after the emergence of permanent commercial talkies.

When scholars have written about sound, they have generalised and not considered the competition within Australia nor the diversity of the companies and individuals involved in the coming of the talkies. Australian sound companies, freelance inventors, backyard tinkerers, businessmen and manufacturers all challenged what appeared initially to be a dominant American influence on sound technology. Australians were as obsessed with modernity and sound's capabilities as pioneers across the globe. As seen in this chapter, individual and at times common efforts to modernise the Australian industry involved a series of proactive and reactive strategies, leading to a virtual war of words in the trade advertisements surrounding the plethora of wares heralded to the industry.

Clearly, this was a period in which 'locally made' was a sign of a complex collaboration among Australian business interests as the world changed because of ambivalence. One might say there was a battle for modernity and standardisation that Raycophone, Australtone, Markophone, Auditone and many other sound systems promoted in articles and advertisements in the popular and trade press during this time. These alternative systems helped wire thousands of cinemas in suburban and rural locations, which the big American electric firms could not reach or simply had no interest in reaching—symbiotically encouraged by exhibitors keen to explore the novelty and profitability of sound.

When left to their own devices, Australians helped create fertile ground for the exhibition and possibly the surprising success of domestic sound films. Raycophone, Australtone, Markophone and Auditone equipment fulfilled a demand for modern Australian sound technology while stimulating the rise of other systems functioning as alternatives to them and Western Electric. In these ways, Allsop, the Tighes, Snider, Betts and Noad as well as Ward and Hawkins acted as agents of modernity while their systems were indeed a kind of local means for exhibitors to participate in the modernisation of the domestic cinema. This small but energetic group exercised everything they could to be part of the sound revolution—namely, part of modernity and the changing audiovisual landscape of the metropolis.

There was something far more interesting happening during this period than a nation waiting to be 'awakened' by sound and Americans bringing new technology to a culture living in the dark ages. The coming of sound is as much a story about the embracing of ideas of urban modernity as it is about responding to the fears that empire loyalists, teachers, ministers, local officials and women's groups expressed against the possible side-effects US films might have on young Australians. It was partly these fears that developers of local, regional and national Australian sound systems exploited in order to gain as many installation contracts as possible.

There are many lessons that can be learned about maintaining a local presence in an incredibly 'global' world. In Australia's talkie war, American interests demonstrated and dictated where, when and how the battles would be fought. All the companies had to meet American terms or go down in defeat. Yet the war rhetoric ignores the fact that Raycophone (and others) did innovate, adopt, thrive and survive—not by defeating the Americans but by assimilating them, by making them Australian rather than shaping the Australian cinema in a solely American way.

Acknowledgements

Archival research was made possible through a URC-New Researcher Grant endorsed by the Faculty of Arts and the Centre for Asia Pacific Social Transformation Studies (CAPSTRANS) at the University of Wollongong. The author thanks the dedicated and very helpful archivists at the National Archives of Australia, National Library of Australia, National Film and Sound Archive, State Records of South Australia, NSW Public Records Office and Australian Patent Office in Canberra and Melbourne. Additional thanks go to Joy Damousi and Desley Deacon for extending an invitation to join this exciting project and to Kate Bowles and Chris Barker for their helpful feedback and continuing collegial support. An earlier version of this chapter appears as 'Projecting Sounds of Modernity: The rise of the local "talkie" technology in the Australian Cinema, 1924–32', *Australian Historical Studies*, vol. 35, no. 123, April 2004, pp. 54–83.

ENDNOTES

[1] See Nowell-Smith, G. and S. Ricci 1998, *Hollywood and Europe: Economics, Culture, National Identity, 1945–95*, British Film Institute, London; Higson, A. and R. Maltby (eds) 1999, *'Film Europe' and 'Film America': Cinema, Commerce and Cultural Exchange, 1920–39*, University of Exeter Press, Exeter.

[2] Between June and July 1930, the German Tobis-Klangfilm and American Western Electric-ERPI and RCA Photophone firms assembled at the Paris Picture Sound Conference and deliberated on how to carve up global sales territories for their sound technologies. These particular big electric companies researched and developed sound applications in the first instance, providing essential manpower and hardware for the experimentation, diffusion and eventual wide-scale adoption of sound. The 'Paris Agreement', which was signed on the last day of this conference, divided the world into three sales zones: an exclusive German territory, an exclusive US territory (which included Australia and New Zealand) and a neutral territory. Detailed conference minutes are held in the Academy of the Motion Picture Arts and Sciences Library Archives, MPAA General Correspondence Files, MF Roll #1, 1929–30.

[3] 'Sound projection technology' here refers specifically to the two basic film formats that dominated the US and European film industries at the time. Sound-on-disc systems used gears and chains to synchronise a phonograph (playing gramophone records) with an existing silent projector. Sound-on-film equipment used (microscope) lenses and special patented lamps to read and amplify audio signals, which appeared as sound waves along the edge of the film. Apparently, the tonal qualities of the disc format were superior to those of sound-on-film, however, discs often skipped and caused synchronisation problems. Sound companies continued to improve both formats in order to hedge their product lines (market share) if and when one method became the sole industry standard. Until about 1933 in Australia, disc systems were as popular as sound-on-film (or at least in the larger cinemas wired with dual systems—under pressure from Hollywood distributors—in order to maximise access to the widest number and types of sound films).

[4] Western Electric opened offices in Sydney in January 1926 in order to 'modernise' Australia with American telephone and electrical technology. In early 1927, Western Electric's sales and service company, Electrical Research Products Inc. (ERPI), filed Australian patents for its 'Talking Motion Picture System', which offered sound-on-disc, sound-on-film or dual configurations. The company was ready to dominate the Australian cinema—as Australians would soon find out. See 'Western Electric Company (Australia) Pty Ltd Memorandum of Association and Special Resolution Adopting New Articles of Association, 11 January 1926', Defunct Company Records, File #934-F2255, NSW Public Records Office, Sydney; and 'Electrical Research Products Inc., Application For Patent on Behalf of a Company as Assignee of the Actual Inventor, 18 May 1927', Public Inspection Correspondence Records, File #7947/27, Australian Patent Office, Canberra.

[5] Matthews, J. J. 2005, *Dance Hall & Picture Palace: Sydney's Romance With Modernity*, Currency Press, Sydney, p. 15. Matthews offers a fascinating account of Australia's engagement with modernity between the economic recessions of the 1890s and the 1930s.

[6] The only exception in the race to screen commercial talkies in Australia at the end of 1928 seems to have been a mobile disc system known as the Han-a-phone, which was imported from the United States and promoted as the 'latest marvel of the age' and 'leader of the new era of sound'. The licence-holders of the Han-a-phone targeted cinemas in country towns, although the project failed after mid-1929 due to difficulties finding trained operators. See *Film Weekly*, 8 November 1928, p. 8; and *Film Weekly*, 13 June 1929, p. 16.

[7] For the most part, Western Electric equipment was rented or licensed to exhibitors and cinema chains. Contracts were negotiated on an individual ad hoc basis in order to extract the highest potential revenues. Prices, if advertised at all, usually evaded the duty, transportation, maintenance and disc replacement costs, making this particular US technology cost prohibitive for most suburban showmen. See *Everyones*, 14 November 1928, p. 7. In June 1929, *Film Weekly* reported that the King's Cross cinema in Darlinghurst, Sydney, had installed Western Electric equipment at a rough cost of £6,000, though the company was promising to lower costs some time in the near future. See *Film Weekly*, 27 June 1929, p. 3.

[8] See Gomery, D. 1980, 'Hollywood Converts to Sound: Chaos or Order?', in E. W. Cameron (ed.), *Sound and the Cinema: The Coming of Sound to American Film*, Redgrave, South Salem, New York, pp. 24–37; and Gomery, D. 1985, 'The Coming of Sound: Technological Change in the American Film Industry', in T. Balio (ed.), *The American Film Industry*, R evised edition, University of Wisconsin Press, Madison, pp. 229–51.

[9] Along with the MPDA of Australia, the Film Renters' Association was dominated by American interests. All but one member of the association—Greater Australasian Films Ltd—worked for American

distribution exchanges. See Tulloch, J. 1982, *Australian Cinema: Industry, Narrative, and Meaning*, George Allen and Unwin, Sydney.

[10] Collins, D. 1987 , *Hollywood Down Under—Australians at the Movies: 1896 to the Present Day*, Angus & Robertson, North Ryde, NSW, p. 77.

[11] 'De Forest Phonofilms (Australia) Ltd Certificate of Incorporation and Memorandum and Articles of Association, 22 August 1925', Defunct Company Records, File #9756, NSW Public Records Office, Sydney.

[12] See, for instance, *Film Weekly*, 21 April 19 27, p. 13.

[13] See Yecies, B. 2005 , 'Transformative Soundscapes: Innovating De Forest Phonofilms Talkies in Australia', *Scope: An Online Journal of Film Studies*, no. 1, available from www.nottingham.ac.uk/film/journal (accessed February 2005).

[14] See *Film Weekly*, 8 June 1927, p. 9. All sound-on-film systems were seen as an infringement on De Forest's US patents. Exhibitors installing Western Electric (Fox Movietone) sound-on-film were directly implicated, though the Western Electric (Warner Vitaphone) disc system was not considered an infringement because it was significantly different to the Phonofilms format.

[15] The warnings could have been a bluff because at the time the company's legal rights were still being decided in US courts. In any case, this was one of the only ways in which the Australian De Forest Phonofilms Company could protect its market share from competition because patent infringement suits were one of the legal weapons used to protect rights to collect royalties.

[16] *Everyones*, 28 November 19 28, p. 7; *Film Weekly*, 29 November 1928, p. 10.

[17] *Everyones*, 2 January 19 29, p. 6.

[18] *Everyones*, 9 January 19 29, p. 4.

[19] *Everyones*, 14 November 19 28, p. 7.

[20] Archival research indicates that installation contracts (also known as wiring contracts) were agreements that exhibitors and cinema owners made with sound companies and/or individual engineers to wire their cinemas with sound equipment. Contracts specified how a company would convert a cinema to sound, including the particular brands of sound equipment used and service and maintenance fees. See also *Everyones*, 8 May 19 29, p. 6; *Everyones*, 5 June 1929, p. 6; *Argus*, 5 July 1929, p. 11; *Film Weekly*, 11 July 19 29, p. 16.

[21] See, for example, Harden, F. 1990, 'Timeline: 1895–1930', *Cinema Papers*, no. 78, pp. 20–5; Reade, E. 1975 , *The Australian Screen*, Lansdowne Press, Melbourne, p. 143; Allard, A. 1989, 'Grand Gala of Gab (1928–1939)', in I. Bertrand (ed.), *Cinema in Australia: A Documentary History*, University of New South Wales Press, Kensington, NSW, pp. 121–76; Stockbridge, S. 1979 'Monopoly Capitalism: T he Case of the Australian Film Industry', *The Australian Journal of Screen Theory*, v ols 5–6, pp. 17–35; Dermody, S. 1981, 'Rugged Individualists or Neocolonial Boys? The Early Sound Period in Australian Film, 1931/2', *Occasional Papers in Media Studies—Faculty of Humanities and Social Sciences*, New South Wales Institute of Technology, Broadway. Among these studies, few systems are named or credited for their part in trying to break the American monopoly—confirming Hollywood's dominance.

[22] Shirley, G. and B. Adams 1989 , *Australian Cinema: The First Eighty Years*, Currency Press, Hong Kong, p. 104. The best known of these was Ray Allsop's Raycophone, but the Auditone system is mentioned in the two paragraphs dedicated to the Raycophone.

[23] Tulloch, *Australian Cinema* , p. 17.

[24] Walsh, M. 1999 , 'The Years of Living Dangerously: Sound Comes to Australia', in D. Verhoeven (ed.), *Twin Peaks*, Australian Catalogue Co., Melbourne, p. 69.

[25] Ibid., p. 72.

[26] Geduld, H. M. 1975 , *The Birth of the Talkies: From Edison to Jolson*, Indiana University Press, Bloomington, p. 252.

[27] *Film Weekly*, 28 March 19 29, p. 10. The editors of *Film Weekly* proclaimed confidently that Allsop's system was 'no hot air'.

[28] See 'Local Talkie Has Secret Try-Out', *Everyones*, 6 February 1929, p. 6; and 'Talking Films: New Australian Equipment, First Public Demonstration', *Sydney Morning Herald*, 11 June 1929, p. 15.

[29] *Film Weekly*, 17 April 19 30, p. 36.

[30] *Everyones*, 16 June 19 37, p. 4.

[31] Fascinating details concerning Allsop's invention can be found in the following archival documents: 'Raycophone Ltd' , File #17/8897-12589, NSW State Records, Kingswood, Sydney; and 'Raycophone Ltd', File #A461/9-S301/1/3, Australian National Archives, Canberra.

[32] Allsop, R. C. 1970 , 'ABC Radio Interview—Transcript, Sydney, September 1970', in the Allsop Collection, National Library of Australia, Canberra, File# MS 421, pp. 14– 15.

[33] Ritchie, J. (ed.) 1993, 'Allsop, Raymond Cottam', *Australian Dictionary of Biography: A–De*, vol. 13, 1940–80, Melbourne University Press, Melbourne, p. 38.

[34] *Everyones*, 6 February 19 29, p. 6.

[35] *Everyones*, 9 January 19 29, p. 4.

[36] *Everyones*, 6 February 19 29, p. 7.

[37] See Tulloch, *Australian Cinema* , p. 62. During this time, according to company records, John Tait was also a large shareholder in Australasian Films Ltd, which was managed by Stuart Doyle (File #17-8851/4723, Australian Archives of NSW, Sydney). Hoyts Theatres' company records for this period also indicate that John Tait was on its board of directors with Frank Thring during the coming of sound (Hoyts File #1 and 2, Melbourne University Archives).

[38] Keith Murdoch's involvement in the motion picture industry continued while supporting Raycophone. In December 1932, Murdoch's *Herald and Weekly Times* financed Melbourne-based Australian Sound Films Ltd, which was established to make news-reel films exclusively for Murdoch's growing newspaper empire. See *Everyones*, 14 October 1931, p. 9.

[39] 'Australian Synchronised Sound Pictures Ltd Memorandum and Articles of Association, 31 August 1929', Defunct Company Records, File #10515/7551, NSW Public Records Office, Sydney.

[40] *Everyones*, 31 July 19 29, p. 6.

[41] Australtone's locally made sound films included songs, violin solos, comedy acts and pianoforte selections performed by Fred Bluett, Hector St Clair, Mabel Barham, Les Rohmer and Bert Corrie. Higgins used recording equipment belonging to the Tighes to shoot the shorts in the Lecture Hall of the Royal Sydney Show Grounds, where the Tighes had constructed a small studio for local productions. The synchronised discs were then pressed by Broadcasters Ltd, Melbourne, which was affiliated with the Sydney company that Allsop worked with. See *Film Weekly*, 18 July 1929, p. 3; and *Film Weekly*, 8 August 19 29, p. 12.

[42] *Everyones*, 9 October 19 29, p. 21.

[43] In comparison, Queensland cinemas had 106 dual and 12 disc-only projectors. Victoria and NSW had the highest number of cinemas wired for sound—most of which were dual projectors. Western Electric and Raycophone made most of the installations in NSW—therefore, cinemas in NSW experienced the most pressure to adopt alternative sound systems that emulated standards set by Western Electric and Raycophone. See *Everyones*, 10 December 1930, p. 14.

[44] *Film Weekly*, 8 August 19 29, p. 21.

[45] As early as 1918, E. L. Betts believed strongly in marrying 'appropriate and well played' music with pictures in order to create a film's full effect. See *Theatre Magazine*, 1 February 19 18, pp. 33–4.

[46] Hoyts Theatres' executives created this company to offer its large exhibition chain a cheaper way to wire the majority of its smaller suburban and country cinemas outside Hoyts Theatres' contracts with Western Electric. See 'Markophone Ltd Memorandum and Articles of Association, 3 February 1930', Defunct Company Records, File #17/5742-13203, NSW Public Records Office, Sydney.

[47] 'Auditone Ltd Memorandum and Articles of Association, 28 October 1929' , Defunct Company Records, File #17/5726-12888, NSW Public Records Office, Sydney.

[48] Between 1922 and 19 26, Hawkins was a producer with Sydney-based Sovereign Pictures, where he made educational, scientific and industry films for the NSW government and commercial organisations. He joined Phonofilms Australia when it merged with Sovereign Pictures. See *Everyones*, 8 December 1926, p. 18; and Hawkins, Stanley W. 1927, 'Interview with Walter M. Marks', *Report of the Royal Commission on the Moving Picture Industry in Australia* , The Parliament of the Commonwealth of Australia, Canberra, 29 July 1927, pp. 413–21.

[49] In August 1929, Broadway Theatres Ltd removed Auditone systems from two of its theatres and cancelled pending installation contracts. The company then installed Western Electric equipment at its Acme Theatre. Thus, Western Electric-ERPI had successfully sold its equipment to one of the largest investors and supporters of the Auditone system, and Auditone became one of the systems that exhibitors were throwing on the scrap heap, which Western Electric's advertisements exploited. See *Film Weekly*, 15 August 1929, p. 18; *Everyones*, 11 June 1930, p. 13; and *Film Weekly*, 3 March 1932, p. 32.

[50] Lumenthode and Ward Sound were still operating in mid-1933, however, by 1936, all of the sound companies that Ward and Hawkins had been associated with had ceased to be recognised in *Film Weekly*'s 'Who's Who' supplement list of 'Sound Equipment Companies Operating in Australia' (24 September 19 36, p. 28). A majority of Australia's cinemas had been wired by this time, and few new installations were needed.

[51] *Everyones*, 16 June 19 37, p. 4.

[52] *Film Weekly*, 10 January 19 29, p. 16.

[53] *Everyones*, 4 June 1930, p. 43.

[54] *Everyones*, 26 November 19 30, p. 47.

[55] *Everyones*, 11 December 19 29, p. 106.

[56] *Everyones*, 7 August 19 29, p. 21.

[57] *Film Weekly*, 25 August 19 32, p. 19.

[58] *Everyones*, 21 May 1930, p. 21.

[59] *Film Weekly*, 17 April 19 30, p. 36.

[60] Service was part of a sound company's ability to deliver a quality product as well as its appeal to exhibitors. Western Electric and RCA established service centres throughout Australia in order to better provide clients with general maintenance and equipment upgrades. Western Electric's mandatory 10-year fee-based service contracts meant, however, that local companies that did not demand a service contract could gain a market advantage. RCA most likely became a popular alternative American system—at least in the Australian market—because it offered a free service plan. Furthermore, companies such as Markophone, which promoted its 'foolproof' equipment, also appeared popular initially with exhibitors who did not want to worry about expensive and time-consuming repairs.

[61] According to *Film Weekly*, 58 Australian films were produced and released commercially between 1930 and 1939. See the ' Film Weekly Motion Picture Directory, History of Australian Production', *Film Weekly* , 1969– 70, Sydney, pp. 67– 72.

[62] *Argus*, 14 June 1926, p. 10.

[63] *Argus*, 11 June 1929, p. 11.

[64] *Argus*, 7 August 1929, p. 7.

[65] *Argus*, 31 October 1929, p. 8.

[66] *Argus*, 15 April 1930, p. 9.

[67] *Argus*, 18 October 1930, p. 6.

[68] In June 1926, legislators in Victoria commented that 'the amount of American films coming into Australia was acting to the detriment of the British Empire and was having an ill effect on the young people of Australia'. See *Argus*, 14 June 1926, p. 10. In the same month, after returning from an education conference in Canada, the Australian Director of Education, S. H. Smith, claimed that 'we need legislation to enable our censors to exclude all talking pictures which disregard the cannons of pure speech as practiced amongst the educated classes in British communities'. See *Sydney Morning Herald*, 10 June 1929, p. 11.

Authors

Alan Atkinson is ARC Professorial Fellow at the University of New England in New South Wales. He is the author of *The Europeans in Australia, Volume 1* (Oxford, 1997) and *The Europeans in Australia, Volume 2* (Oxford 2004); *The Commonwealth of Speech* (Australian Scholarly Publishing, 2002), *The Muddleheaded Republic* (1993) and *Camden: Farm and Village Life in Early New South Wales* (Oxford, 1988). He is currently working on volume three of *The Europeans in Australia*.

Diane Collins is Associate Dean (Teaching and Learning) at the Sydney Conservatorium of Music, where she has been course director and lecturer in the Historical and Cultural Studies programs since 1995. She has published *Sounds from the Stables: The Story of Sydney's Conservatorium of Music* (Allen and Unwin, 2001), *Hollywood Down Under: Australians at the Movies 1896 to the Present Day* (Angus & Robertson, 1988) and, with Ina Bertrand, *Government and Film in Australia* (Australian Film Commission and Currency Press, 1981). She is currently working on a book entitled *Acoustic Journeys: Explorations in the Aural History of Australia*, which deals with aspects of Australian auditory history in the colonial period and early twentieth century.

Joy Damousi is Professor of History and Head of the School of Historical Studies at the University of Melbourne. Her recent areas of publication include memory and the history of emotions—themes she explored in two previous publications, *The Labour of Loss: Mourning, Memory and Wartime Bereavement in Australia* (Cambridge, 1999) and *Living with the Aftermath: Trauma, Nostalgia and Grief in Post-War Australia* (Cambridge, 2001), and in the collection of essays edited with Robert Reynolds, *History on the Couch: Essays in History and Psychoanalysis* (Melbourne University Press, 2003). Her latest publication is *Freud in the Antipodes: A Cultural History of Psychoanalysis in Australia* (University of NSW Press, 2005). One of her current research projects is 'Elocution Lessons: A history of elocution and the auditory self in Australian cultural life'.

Desley Deacon is Professor of Gender History and Head of the History Program of the Research School of Social Sciences at The Australian National University and president of the Australian Historical Association. She is the author of *Elsie Clews Parsons: Inventing Modern Life* (University of Chicago Press, 1997) and the forthcoming *Mary McCarthy: Four Husbands and a Friend* (University of Chicago Press, 2008) (the friend is Hannah Arendt). The lives of these women—and her own experiences teaching for 16 years in the American Studies Department at the University of Texas at Austin—stimulated her long-standing interest in internationalism. Her interest in the history of the voice was provoked by initial work on her new project 'Judith Anderson 1897–1992: voice and emotion in the making of an international star'.

James Donald is Dean of the Faculty of Arts and Social Sciences and Professor of Film Studies at the University of New South Wales, having taught previously at the Open University and Sussex University in England, and at Curtin University of Technology in Western Australia. He is author of *Sentimental Education: Schooling, Popular Culture and the Regulation of Liberty* (Verso, 1992) and *Imagining the Modern City* (Athlone Press/University of Minnesota Press, 1999), co-author of the *Penguin Atlas of Media and Information* (Penguin, 2001) and editor of a dozen books on the cinema, media, education and social theory.

Bridget Griffen-Foley is an ARC Queen Elizabeth II Fellow and the Director of the Centre for Media History at Macquarie University. Her publications include *Party Games: Australian Politicians and the Media from War to Dismissal* (Text, 2003) and she is now writing a history of commercial radio in Australia.

Bruce Johnson, former Professor in English, University of New South Wales, is now Adjunct Professor of Contemporary Music Studies, Macquarie University; Honorary Professor of Music, University of Glasgow; and Visiting Professor in Cultural History, University of Turku, Finland, where he is also on the committee steering the International Institute for Popular Culture. He has published on popular music, acoustic ecology and Australian studies, including *The Oxford Companion to Australian Jazz* (Oxford University Press, 1987) and *The Inaudible Music: Jazz, Gender and Australian Modernity* (Currency Press, 2000). He is currently co-authoring a book on music and violence for Ashgate Press and editing a collection on sound and pornography for Equinox. He has also been active in radio broadcasting, arts administration—including in the establishment of the government-funded Australian Jazz Archives—and is a jazz musician with international experience in recording, touring and festival work.

Associate Professor **Peter Kirkpatrick** teaches in the English, Text and Writing Program at the Penrith campus of the University of Western Sydney. He writes cultural history and literary criticism, and his publications include, with Jill Dimond, *Literary Sydney: A Walking Guide* (University of Queensland Press, 2000) and *The Sea Coast of Bohemia: Literary Life in Sydney's Roaring Twenties* (University of Queensland Press, 1992; rev. edn API Network, 2007). He also edited *The Queen of Bohemia: The Autobiography of Dulcie Deamer* (University of Queensland Press, 1998) and *Ronald McCuaig: Selected Poems* (Angus & Robertson, 1992).

Marilyn Lake holds an Australian Professorial Fellowship and Personal Chair in History at La Trobe University. From 2001 to 2002, she was Chair of Australian Studies at Harvard University and she has also held Visiting Professorships at the University of Western Australia, Stockholm University, the University of Sydney and The Australian National University. She has published widely on questions of gender, citizenship, sexuality, race and nationalism. Her biography *Faith: Faith Bandler, Gentle Activist* (Allen and Unwin, 2002) won the HREOC award for non-fiction in 2002; her most recent books are *Memory, Monuments and Museums* (Melbourne University Press, 2006) and, co-edited with Ann Curthoys, *Connected Worlds: History in Transnational Perspective* (ANU Press, 2006). Her next book, co-authored with Henry Reynolds, *Drawing the Global Colour Line: White Men's Countries and International Campaigns for Racial Equality*, will be published jointly by Cambridge University Press in the United Kingdom and Melbourne University Press in Australia.

Bruce Moore is Director of the Australian National Dictionary Centre at The Australian National University. His recent publications include *Gold! Gold! Gold! The Language of the Nineteenth-Century Australian Goldfields* (Oxford University Press, 2000), *The Australian Concise Oxford Dictionary*, 4th edn (Oxford University Press, 2003), *The Australian Oxford Dictionary*, 2nd edn (Oxford University Press, 2004), *Australian Aboriginal Words in English*, 2nd edn, with R. M. W. Dixon, W. S. Ramson and Mandy Thomas (Oxford University Press, 2006) and *The Australian Oxford Paperback Dictionary*, 4th edn, with Frederick Ludowyk (Oxford University Press, 2006). He is currently writing a book on the history of Australian English and editing the second edition of *The Australian National Dictionary*.

Brian Yecies is Lecturer in Media and Cultural Studies at the University of Wollongong, Australia. His research on film policy and industry in colonial Korea and post-colonial South Korea and the coming of sound to the Australian cinema (1924–39) has appeared in the *Journal of Korean Studies*, *Korea Observer*, *Asian Cinema*, *Yonsei Institute of Media Arts New Korean Cinema Series*, *Scope: An Online Journal of Film Studies*, *Screening the Past* and *First Monday: Peer-Reviewed Journal* on the Internet. In 2003, he received an invaluable research grant from the Asia Research Fund and, in 2005, he was a Korea Foundation Research Fellow at the Korean National University of the Arts.

Index

www.ingramcontent.com/pod-product-compliance
Lightning Source LLC
Chambersburg PA
CBHW061246270326
41928CB00041B/3437